Frogs: *The Animal Answer Guide*

Frogs
The Animal Answer Guide

Mike Dorcas and Whit Gibbons

The Johns Hopkins University Press   Baltimore

© 2011 The Johns Hopkins University Press
All rights reserved. Published 2011
Printed in the United States of America on acid-free paper
9 8 7 6 5 4 3 2 1

The Johns Hopkins University Press
2715 North Charles Street
Baltimore, Maryland 21218-4363
www.press.jhu.edu

Library of Congress Cataloging-in-Publication Data

Dorcas, Michael E., 1963–
   Frogs : the animal answer guide / Mike Dorcas and Whit Gibbons.
     p.   cm.
   Includes bibliographical references and index.
   ISBN-13: 978-0-8018-9935-5 (hardcover : alk. paper)
   ISBN-10: 0-8018-9935-4 (hardcover : alk. paper)
   ISBN-13: 978-0-8018-9936-2 (pbk. : alk. paper)
   ISBN-10: 0-8018-9936-2 (pbk. : alk. paper)
   1. Frogs. I. Gibbons, J. Whitfield, 1939– II. Title.
QL668.E2D66 2011
597.8'9—dc22   2010042484

A catalog record for this book is available from the British Library.

Special discounts are available for bulk purchases of this book. For more information,
please contact Special Sales at 410-516-6936 or specialsales@press.jhu.edu.

The Johns Hopkins University Press uses environmentally friendly book
materials, including recycled text paper that is composed of at least 30 percent
post-consumer waste, whenever possible. All of our book papers are acid-free,
and our jackets and covers are printed on paper with recycled content.

Be kind and tender to the Frog,
And do not call him names,
As "Slimy skin," or "Polly-wog,"
Or likewise "Ugly James,"
Or "Gap-a-grin," or "Toad-gone-wrong,"
Or "Bill Bandy-knees":
The Frog is justly sensitive
To epithets like these.

No animal will more repay
A treatment kind and fair;
At least so lonely people say
Who keep a frog (and, by the way,
They are extremely rare).

—Hilaire Belloc

Contents

Contents

Acknowledgments

We are grateful to the numerous staff and students at Davidson College and the Savannah River Ecology Laboratory for their encouragement and assistance in writing this book. In particular, we thank Margaret Wead for assisting in a variety of ways with the details of manuscript preparation. We thank Kimberly Andrews, Rick Bauer, Kurt Buhlmann, Adrien Domske, Evan Eskew, Shawna Foley, Alastair Freeman, Cris Hagen, Tom Luhring, David Millican, Tony Mills, Steve Price, Sean Poppy, David Scott, Charlotte Steelman, Brian Todd, Tracey Tuberville, J. D. Willson, and Lynea Witczak, all of whom were helpful in making suggestions and providing insights from their own experiences with frogs that improved this book. Lynea Witczak was particularly helpful by compiling the bibliography and assisting with the images. Patricia West of American University provided library assistance. Michael Gibbons, Cris Hagen, and J. D. Willson helped by tracking down or providing information we needed for the book.

We especially thank several frog specialists for their willingness to provide comments on particular questions, although we assume full responsibility for the final answers. The following provided comments that substantially improved the manuscript: Michael Gibbons, Cris Hagen, Tony Mills, and J. D. Willson. Joe Mendelson of Zoo Atlanta was particularly gracious to answer our repeated queries about his research. We appreciate the help of numerous photographers who generously provided images for us to use in the book. These photographers include Graham Alexander, Jonathan Campbell, Dante Fenolio, Mike Gibbons, Chris Gillette, Cris Hagen, Aubrey Heupel, Pierson Hill, H. B. Hines DERM, Vic Hutchison, Trip Lamb, Victor Lamoureux, Thomas Luhring, John Mackway, Shannon Pittman, Zbynek Rocek, Leslie Ruyle, Charlotte Steelman, Clyde Sorenson, Wayne Van Devender, John D. Willson, and Robert T. Zappalorti. We thank Susan Harris for compiling the index and assistance with proofing. Finally, we thank our families for their support, assistance, and understanding while we worked on this book. MD thanks his wife Tammy, and his children Taylor, Jessika, and Zachary. JWG thanks his wife Carol, and his children Laura, Jennifer, Susan, and Michael.

Introduction

Both of us grew up as herpetologists. We may not have realized it until later, but when we were young, we somehow acquired (or were born with?) a fascination for wildlife and, in particular, amphibians and reptiles. Such a fascination, one that requires exploration and further understanding, is a prerequisite for being a herpetologist. Thus, neither of us can pinpoint a moment in our lives when we "became" herpetologists. Instead, it is as if we always were. Most herpetologists have similar stories; that is, they grew up that way and have somehow been able to take their fascination and excitement while learning about frogs, turtles, and snakes and turn it into their life's work and a career. Such fascination and interest by most herpetologists results in a remarkable dedication or, some might say, obsession regarding their work. We must be careful not to forget how fortunate we are to be able to make a living doing what we are most passionate about.

Like many herpetologists, we both were initially interested in snakes, but our interest in other groups of amphibians and reptiles increased dramatically as we learned more about their biology. One of our passions was frogs and toads, and each of us has had exciting and memorable experiences looking for frogs at night during large breeding choruses. As we became more experienced herpetologists, we learned about the fascinating life history variation among frogs and toads. Regarding reproduction, frogs nearly "do it all." That is, think of any conceivable way to reproduce and a frog somewhere has figured it out beforehand. Both of us have conducted considerable research on frog biology and conservation and have had the opportunity to work with colleagues who are experts in the field. JWG has conducted several long-term studies on frogs and has studied them with numerous colleagues on the Savannah River Site in South Carolina for more than 40 years. MD began research on frogs in the early 1990s in Idaho and has worked collaboratively with students on many frog-related research projects in the years that followed.

The alarming and catastrophic declines in frog populations worldwide have led to a dramatic increase in the number of people interested in frogs and in the amount of research conducted on frogs. Clearly, we cannot adequately discuss all of this research in such a book. Instead, our goal is to answer questions posed in a way that allows the reader to gain insight into frogs and toads and to provide the reader with information that will raise their awareness regarding the importance of frogs and toads as part of our natural world.

Frogs: *The Animal Answer Guide*

Chapter 1

Introducing Frogs

What are frogs?

Frogs are members of the class Amphibia, often called amphibians, which includes all frogs and toads in addition to two other less-familiar groups of animals. Frogs and toads are vertebrates—animals with backbones—which are generally recognizable and not easily confused with other animals. People rarely, if ever, say, "Is that a frog?" Instead, they are interested in identifying specific frogs or toads. Frogs and toads differ from two other groups of amphibians, salamanders and caecilians, because they lack a tail as an adult. Thus, they are placed in the order Anura, which means "no tail." Scientists often refer to frogs and toads, collectively, as *anurans*. In addition to lacking a tail, most frogs and toads have long hind limbs modified for jumping or, at least, hopping. Frogs and toads are well known for their reproductive behavior, which, for most species, includes loudly calling at certain times of the year to attract mates. Listening to frogs calling around a wetland or swamp is a relaxing pastime. Most frogs and toads also lay eggs, and many species, including those most familiar to people, have an aquatic larval stage usually referred to as a *tadpole*. Frogs are often used as excellent examples of metamorphosis: tadpoles slowly develop legs, lose their tail, and crawl or hop onto land. Although all frogs and toads have the same basic body plan, their varying body forms and other aspects of their biology make frogs and toads an amazingly diverse and significant group of animals. Of the three main groups of amphibians, frogs and toads comprise approximately 90 percent of all known species. In addition, they are the most widespread group of amphibians in terms of geographic distribution and, in many regions, represent the only type of amphibian present.

What is the difference between frogs and toads?

People become confused about the differences between frogs and toads and how to tell one from the other. Scientists consider all frogs and toads to be members of the order Anura, or the "tailless" amphibians, and often refer to them as "anurans." But the terms *frog* and *toad* are not necessarily restricted to specific groups of anurans. In general, frogs are considered to be anurans with relatively smooth skin, long hind legs designed for jumping long distances, and heavily webbed toes for swimming or adhesive toepads for climbing. Frogs often tend to be more aquatic than toads and include familiar species, such as bullfrogs and leopard frogs. Some frogs, especially treefrogs, are much more arboreal (tree-climbing). Toads are usually considered anurans that have relatively short legs designed for hopping shorter distances, little to no webbing between their toes, and limited ability to climb. Their bodies are typically stouter than frogs and have drier skin, often covered with bumps that resemble warts. When amphibian biologists use the term *true frogs*, they are usually referring to members of the family Ranidae, which includes northern leopard frogs (*Rana pipiens*), pickerel frogs (*Rana palustris*), bullfrogs (*Rana catesbeiana*), or the common frog (*Rana temporaria*) found in Europe. The term *true toads* refers to members of the family Bufonidae, which includes Fowler's toads (*Bufo fowleri*), the common toad (*Bufo bufo*) of Europe, and the tiny oak toad (*Bufo quercicus*). However, Ranidae and Bufonidae are only 2 of more than 30 anuran families, and most anurans are not members of either of these families. Consequently, most scientists who study amphibians refer to the entire group as anurans. Among scientists and laypeople, it is acceptable to refer collectively to all anurans as *frogs*, reserving *toads* for less-aquatic anurans with relatively short legs and relatively dry skin. In some parts of the United States, the term *toad* is applied not only to true toads but also to the spadefoot toads in the family Scaphiopodidae and the narrowmouth toads in the family Microhylidae.

What other types of amphibians are there?

Although frogs and toads (anurans) represent the most abundant and diverse group of amphibians on earth, two other primary amphibian groups are extremely interesting and, in some parts of the world, represent important components of the ecosystems in which they live. One group of amphibians other than anurans is salamanders. If frogs and toads are the tail-

Frogs: The Animal Answer Guide

A typical toad, such as this American toad (*Bufo americanus*), has a stout body, short legs, and rough, dry skin.
Courtesy John D. Willson

less amphibians, then salamanders are tailed amphibians. All salamanders are in the order Caudata, which means *tail*. Another name frequently used for salamanders is the order Urodela. Salamanders are sometimes referred to as *caudates*, or *urodeles*. Most have two pairs of legs that are not modified for jumping as are the legs of frogs and toads. However, species in one group of salamanders known as sirens have only a pair of front legs and no hind limbs. Unlike frogs and toads, most salamanders do not vocalize. Salamanders are not as widespread as frogs and toads in their worldwide distribution. Salamanders' centers of diversity are eastern North America and Central America, but some members of the group are found throughout Europe and Asia. A limited number of salamander species are also found throughout parts of South America. Salamanders do not occur in sub-Saharan Africa or in Australia. No amphibians are found in Antarctica.

Salamanders have an amazing diversity of species, and those in the largest family, Plethodontidae, are lungless. As adults, they respire through their skin. Many salamanders are fossorial, often burrowing or hiding beneath logs, rocks, and other objects, and some species are found only in caves. Most salamanders have internal fertilization of their eggs. In most species, the male deposits a packet of sperm called a *spermatophore*, which the female picks up with her cloaca, allowing for internal fertilization of her eggs. Many species have an aquatic larval stage like frogs and toads, but larval salamanders typically have well-developed legs and external gills that resemble flimsy feathers sticking out from the sides of their head. Some species of salamanders transform into an adult form but remain aquatic and are able to reproduce while retaining larval characteristics. Such a phe-

The southern leopard frog (*Rana sphenocephala*) is one of the "true frogs," with smooth, moist skin and stout legs for jumping. Courtesy John D. Willson

nomenon is known as *neoteny*, or *paedomorphosis*. Salamanders are unique among vertebrates in their ability to regenerate lost limbs or a lost tail.

The third major group of amphibians is the caecilians. Caecilians are blind, limbless, worm-like amphibians that live primarily in tropical regions. They are placed in the order Gymnophiona (sometimes called the Apoda, or Caecilia). Some caecilians are aquatic and resemble eels because flattened parts of their body form fins. Caecilians have rings, or annuli, that encircle their body along their entire length and deeply embedded scales within these annuli. They have short tails and rather blunt heads adapted for burrowing. A short, retractable tentacle on either side of the snout is unique among all vertebrates and is used as a chemosensory organ. Their eyes are small and usually appear as little dark spots beneath the skin. Most have only one fully developed lung. Caecilians are tropical and are found in the American, African, and Asian Tropics, as well as on the Seychelles Islands, Sri Lanka, and the Philippines. They are absent from the Tropics of Madagascar and Australia as well as the Temperate Zones.

Why are frogs important?

Frogs and toads are important animals for several reasons. Frogs are primary components of many natural ecosystems and play vital roles as both predators and prey for other animals. Frogs as predators and prey are discussed in detail in chapter 5. Frogs are not only food for animals, but in some areas, they are a common food for humans as well. In many cultures, the legs of frogs, such as the edible frog (*Rana esculenta*) of Europe and the bullfrog (*Rana catesbeiana*), are eaten. Frog legs are a common dish

Treefrogs, such as this map treefrog (*Hypsiboas geographic*), are characterized by long legs with toepads that aid in climbing. Courtesy John D. Willson

in French cuisine and are often eaten in China. In Indonesia, frog legs are eaten as a soup. In the United States, a favorite pastime of many is frog gigging, where frogs (usually bullfrogs) are hunted in marshes and swamps at night and impaled using a long, three-pointed spear, or gig. Frogs are farmed in many parts of the world as a food source and to provide specimens for biology classes, where they are dissected.

Some frogs are important not because they help get rid of pests, but because they are pests themselves. However, in nearly all cases, humans are responsible for introducing frogs into regions where they do not belong and where they cause problems. The cane toad (*Bufo marinus*), a large toad native to Central and South America, has been introduced onto many islands and in Australia, where it was released to control cane beetles. However, cane toads are toxic to most animals that attempt to eat them. Many animal populations in Australia, including snakes and large native lizards, that eat cane toads have declined in numbers because of this introduced and highly toxic prey. The little coqui (*Eleutherodactylus coqui*), a frog native to Puerto Rico, has been introduced to Hawaii, where it endangers native insect populations potentially important for plant pollination. Local Hawaiian residents complain because these little amphibians produce repetitive, high-pitched calls. Control efforts on Hawaii include spraying the frogs with citric acid, which can kill them when the acid is absorbed through the skin. The Cuban treefrog (*Osteopilus septentrionalis*) has invaded much of Florida and because of its size is able to capture and consume many native species of treefrog. Whether it was introduced by humans is debatable; however, the danger it poses to native treefrog populations is real in some areas, and it has recently spread throughout much of the Florida peninsula.

In the western United States, where the American bullfrog has been introduced outside its native range, environmentalists view the species as invasive and detrimental, a predator on native frogs and other aquatic animals.

Frogs and toads are also important because they are indicators of the ecosystems' health. Animals that indicate environmental health are often referred to as *bioindicators*. Frogs and toads can be particularly sensitive to environmental problems such as pollution. Because they have rather permeable skin and because many species have a biphasic life cycle in which they live on land and in the water, they are particularly vulnerable to toxins in the environment. Frogs and toads have even been touted as "canaries in the coal mine" because they tend to succumb to toxins before the toxin's effects on humans are evident. In 2010, Jacob Kerby and his associates tested whether frogs are good bioindicators. They evaluated thousands of tests of the effects of chemicals on frogs and found that, overall, frogs and toads were somewhat tolerant of many chemicals. They concluded that perhaps we should rethink whether frogs and toads are good bioindicators.

Frogs and toads are important to many cultures as symbols of immortality and fertility and are the subjects of many legends and myths. Frogs are celebrated characters in television shows and in beer commercials. The fairy tale about the frog prince epitomizes humans' fascination and appreciation for frogs and toads over the centuries. Even if all the previous reasons are discarded, frogs and toads are worthy of appreciation in their own right as components of our natural ecosystems. Frogs and toads are interesting, sometimes beautiful animals that have inspired excitement and fascination in humans for millennia. As children, many of us fondly remember searching for and finding frogs and the excitement resulting from such adventures. Ideally, such appreciation leads to a sense of responsibility and a conservation ethos that respects frogs and toads.

Where do frogs live?

Frogs and toads live on every continent except Antarctica. They are also absent from Greenland and many other oceanic islands but are found on many islands, including Madagascar, many of the Caribbean islands, Indonesia, and the Seychelles. Although the greatest concentration of frogs is in the Tropics, where warm and often wet climate provide ideal habitats for many species, frogs and toads are not confined to tropical regions. Some frog species, such as the wood frog (*Rana sylvatica*), are found above the Arctic Circle. Other species are found in some of the driest deserts on earth.

In general, frogs are found in nearly all terrestrial and freshwater aquatic habitats. Frog and toad species are found in and around lakes, ponds,

Spadefoot toads (*Scaphiopus holbrookii*) can remain dormant underground at low temperatures for months or even years without eating by lowering their metabolic rates.

Courtesy John D. Willson

swamps, wetlands, and mountain streams. At least one species, the crab-eating frog (*Fejervarya cancrivora*), a native of Southeast Asia, inhabits mangrove swamps and saltwater marshes, where the water is brackish. Several species of frogs are found at extremely high elevations in different parts of the world. Frogs of the genus *Telmatobius* reach close to or above 12,000 feet in Peru, and the Ladakh toad (*Bufo latastii*), native to Pakistan and India, has been reported to occur near 10,000 feet. In the United States, boreal chorus frogs (*Pseudacris maculata*) have been observed breeding at elevations in Colorado above 12,200 feet. Alpine frogs in the genus *Scutiger* have been reported by some amphibian biologists at elevations above 16,000 feet in the Himalayas.

Many species of frogs from multiple families are arboreal, or tree-dwelling. There is even one species of toad that will climb into bushes and small trees. Some tropical species are so tied to living in trees that they even lay their eggs in little puddles of water at the base of bromeliads. The eggs hatch into tadpoles that metamorphose high in the treetops. Many species of frogs are fossorial and spend much time buried underground. Some spadefoot toads (family Scaphiopodidae) can spend months or even years buried belowground, waiting for the right conditions to emerge and breed.

How many kinds of frogs are there?

More than 5,600 living species of frogs and toads exist and are classified by amphibian biologists into more than 30 taxonomic families. According

The diminutive forest rain frog (*Breviceps sylvestris*) is found only in northeastern South Africa. The female of this species is known to attend her terrestrial nest and the eggs hatch directly into tiny froglets.

Courtesy Graham Alexander

to a 2008 estimate, the total number of distinct species is 5,645, and the number of recognized frog and toad families is 34, but the opinions and interpretations of different experts vary regarding the exact numbers of each. Some families consist of hundreds of species, whereas others have no more than a handful, and two families have only one species.

The largest group of frogs recognizable to most people is treefrogs, which belong to the family Hylidae, with more than 850 species described from the Americas, Europe, Southeast Asia, and Australia. Many have expanded toe tips that aid in climbing and long thin legs for jumping. A completely different family of frogs known as the Asian treefrogs (family Rhacophoridae), which occur in Africa, Asia, and Indonesia, has almost 300 species. These are similar in appearance and behavior to other treefrogs but are not closely related.

Another large and familiar group is the *true frogs*, known as the family Ranidae, and comprising more than 830 species. The most familiar ones are similar-looking species with robust bodies and strong hind legs and include the well-known bullfrog, several kinds of leopard frogs of North America, and the common frog of Great Britain and throughout much of Europe. Also included in the Ranidae are the rare gopher frogs of southeastern United States.

One of the most familiar and largest groups of toads throughout most of the globe is the family Bufonidae, with more than 500 species. Members of this family are native to all warm continents except Australia. However, the tropical American cane toad has been introduced and is now established, albeit unwelcome, in many parts of Australia. The species of toads

Frogs: The Animal Answer Guide

differ among and within countries, but the squat brown insect-eaters, commonly seen in gardens and backyards in most U.S. states, are also found in regions as diverse as Great Britain, China, India, and Argentina.

Two species of frog are distinctive as the only species in their family. The rare burrowing toad (family Rhinophrynidae) of southern Texas ranges through Mexico and Central America to Costa Rica. The purple frog (family Nasikabatrachidae) is also a rare burrowing species from India. The family Ascaphidae has only two species, the tailed frogs of the northwestern United States and Canada, as does the family Rhinodermatidae and the mouth-brooding frogs of Chile. Ten families of frogs have fewer than a dozen species, and 12 families have 25 to 200 species. Aside from those mentioned earlier, the largest families of frogs are the Microhylidae with 434 species, Leptodactylidae with 498, and Strabomantidae with 536. Species in the latter two families are found predominately in Central and South America. Species in the Microhylidae, to which the narrowmouth toads of the United States belong, are more widespread and are found throughout much of the world, including the Americas, Africa, India, southern Asia, Indonesia, and Australia.

The species diversity of frogs and toads is greatest in the tropical regions of South America. For example, 107 species are found in the United States. Russia and Canada each have 24 species, compared with 27 in France, 226 in Australia, and 300 in China. However, all of these are low numbers compared with Ecuador, with 444 species of frogs, Colombia with 678, and Brazil with 776. The greatest concentrations of U.S. species are in the humid southeastern states from the Carolinas to Texas. States with high numbers of native species are Alabama, Georgia, and the Carolinas, with 30 species each. Florida actually has 32 species of frogs, but 3 are introduced exotics. Texas has more species than any other state, with 41, in part because several predominately Mexican species have been found in south Texas. None of the large western states from New Mexico and Colorado to California and Oregon have as many species as any of the coastal southeastern states. No northern states have as many species of frogs as southern ones. For example, North Dakota has only nine species, and Alaska has only three native species. Hawaii is the most depauperate of native frogs, having none. However, a half-dozen species of introduced frogs now live in Hawaii, including species from Japan, Puerto Rico, the U.S. mainland, and tropical America.

How are frogs classified into groups?

Not all frog biologists agree on the proper classification and categorization of frogs and toads. Nearly all frog biologists agree that the classification should match the evolutionary history of the group on which they are working. For example, two species that share a more recent common ancestor than they do with another species should be grouped together. Such a classification is called a *phylogeny*, and a branching tree-like diagram that represents the evolutionary history of frogs (or any animal) is called a *phylogenetic tree*. Various characters are used to determine the classification, or phylogenetic tree, of frogs and toads. Historically, only morphological characters, such as skeleton characteristics or the shape of the frog, were used. Since the 1990s, nearly all frog biologists interested in determining the proper classification of frogs into groups have used characters based on DNA. Genetics has transformed our understanding of frog phylogeny but also has resulted in competing hypotheses regarding the evolutionary history of frogs and toads.

Frogs and toads are typically grouped into families, the numbers of which vary depending on the reference source used, but they generally range between 24 and 50 families. All authorities recognize that treefrogs are in the family Hylidae. A taxonomic family is then composed of one or more genera, and each genus is made up of one or more species. The scientific name of every living thing, including frogs and toads, is composed of its genus and species names. For example, the scientific name of the green treefrog is *Hyla cinerea*—the genus is always uppercase and the species name is always lowercase. In addition, scientific names are always italicized or underlined. All frog families are grouped into three main groups referred to as the ancient frogs (Archeobatrachia), middle frogs (Mesobatrachia), and modern frogs (Neobatrachia).

Current nomenclature for many frog genera and species is in flux. Because this book is intended for a lay audience, we have avoided complicating things unnecessarily and have used traditional generic and species names in the text but have listed proposed synonyms in Appendix A.

What are frogs' closest relatives?

The closest living relatives to frogs and toads are salamanders. That is, all frogs and toads share a more recent common ancestor with all salamanders than they collectively share with any other group of animals, including the other main group of amphibians, caecilians. On the basis of fossil evidence, the ancestors of frogs and toads diverged from salamanders before the end of the Triassic period, approximately 250 million years ago (mya).

The closest living relatives to frogs and toads are the salamanders. The long-tailed salamander (*Eurycea longicauda*) of the southeastern United States is one of the fewer than 600 species of salamanders in the world today.

Courtesy John D. Willson

When did frogs evolve?

The first amphibians are known from fossils of the Devonian period, more than 350 mya and more than 100 million years before the first dinosaurs. The first amphibian fossils unquestionably most similar to modern frogs were from sediments more than 180 mya (see *Prosalirus* in the next section). An earlier fossil (see *Triadobatrachus* in the next section) from the Triassic more than 240 mya had many traits similar to modern frogs but had a tail and no hind legs designed for jumping. Frogs evolved from other amphibian ancestors sometime during the Mesozoic era between the Triassic and Jurassic periods. An interesting consideration is that no fossils of frogs have yet been discovered that lived during the more than 50 millions of years between *Triadobatrachus* and *Prosalirus*.

What is the oldest fossil frog?

Deciding on the oldest known fossil frog depends on how paleontologists define a frog. *Triadobatrachus massinoti* was described from a Lower Triassic formation (more than 240 mya) in Madagascar in 1937 and was long considered by herpetologists to be the earliest known frog. Amphibian authorities agree that the pelvic girdle and skull of the single speci-

Some of the oldest known fossil frogs that were ancestors of modern frogs are believed to have lived between 150 and 180 million years ago. Scientists believe this specimen of *Palaeobatrachus* from Europe was fossilized more than 200 million years ago. Courtesy Zbynek Rocek

men are frog-like features similar to modern frogs, although the species had more vertebrae, including six tail vertebrae. Most experts now consider *Triadobatrachus* to be a closely related amphibian but not a direct ancestor of today's frogs.

The oldest known fossil frog, an indisputable ancestor of modern frogs, was found in northeastern Arizona in 1995. This species, *Prosalirus bitis*, lived in the Early Jurassic, more than 180 mya. Farish Jenkins of Harvard University and Neil Shubin at the University of Chicago described the species based on skeletal parts from three specimens. Although the specimens showed some characteristics of primitive amphibians, the species had modern frog features, including fused tail vertebrae (the urostyle) and elongated hind legs that could be used for jumping. *Prosalirus* were approximately 2 inches long.

Another ancient fossil frog, from the Jurassic in southern South America and a species that may actually have been contemporaneous with *Prosalirus*, is *Vieraella*. However, the age determinations of the deposits are uncertain. *Notobatrachus*, a species described from fossils in the same region, has been estimated to be from sediments at least 158 million and possibly as many as 172 million years old.

Frogs: The Animal Answer Guide

Chapter 2

Form and Function

What are the largest and smallest living frogs?

A single species, the goliath frog (*Conraua goliath*), holds the undisputed record as the largest frog in the world. A native of Cameroon and Equatorial Guinea, the goliath frog inhabits West African rivers, especially near rapids and waterfalls. According to Victor Hutchison, who conducted research on the species, the largest goliath frog ever found was almost 16 inches long (not including the outstretched legs) and weighed more than 6½ pounds. Even the average size of these gigantic frogs in museum collections (7 to 11.8 inches and more than 2 pounds) is greater than the maximum size of nearly all other frogs in the world. Unfortunately, because of their large size, goliath frogs are a favored food item in some areas, which threatens their survival. Because of these and other threats to this impressive species, it is classified as endangered by the IUCN (International Union for Conservation of Nature).

Frogs from other regions can also reach impressive sizes. The biggest frog in the United States and Canada is the bullfrog (*Rana catesbeiana*). The largest individuals reach 8 inches and more than a foot long with legs extended. Their famously huge hind legs have led to their popularity for edible frog legs, but their widespread distribution and abundance have kept them from being endangered in their native range. Two other members of the family Ranidae (the pig frog, *Rana grylio*, and the river frog, *Rana heckscheri*) of North America also have strong back legs and reach large body sizes, although their maximum body length is less than 7 inches.

The body length of the Colorado River toad (*Bufo alvarius*), which is found in the U.S. Southwest (mostly in Arizona) and ranges into Sinaloa

The goliath frog (*Conraua goliath*) of West Africa is the largest frog in the world. Courtesy Vic Hutchison

in Mexico, rivals the bullfrog in body length. The largest individuals reach 7½ inches long. But their hind legs are relatively shorter than bullfrogs' hind legs, and their total lengths with legs outstretched are less. Nonetheless, because of their massive body bulk, the Colorado River toad may grow heavier than any other native U.S. frog. In tropical America, several species are notable for their large sizes. The marine toad (*Bufo marinus*), also known as the cane toad, is arguably the largest toad in the world. Large adult females average 5 to 6 inches in body length, and the maximum size is reportedly more than 15 inches according to some records, with a weight of more than 5 pounds. Two other species, the rococo toad (*Bufo schneideri*) and the yellow cururu toad (*Bufo ictericus*), reach similar or possibly even larger sizes than the marine toad. The South and Central American bullfrog, or smoky jungle frog (*Leptodactylus pentadactylus*), reaches body lengths of more than 7 inches.

Although no one debates that the goliath frog is the largest species in both length and weight, deciding which species is the smallest frog is not so easy. Many tiny contestants can be considered, and the true award for

Frogs: The Animal Answer Guide

The little grass frog (*Pseudacris ocularis*) is the smallest native North American frog. The maximum size of adults is less than one inch.

Courtesy Aubrey Heupel

smallness should go to the species in which the largest individual is smaller than the largest individual of all other species. Determining this is difficult, because who knows when someone might find a slightly bigger individual of the largest known specimen of a rare species at some point in the future. And in a practical sense it does not really matter. With a bit of geographic qualification, certain species would certainly be in the running as the "smallest frog." Most amphibian biologists agree that the frog group with the greatest number of small species is the genus *Eleutherodactylus* in the family Leptodactylidae (or Eleutherodactylidae). But at least three other families (Brachycephalidae, Microhylidae, and Sooglossidae) have tiny representatives as well. Each of these families has a species whose typical adult size is around 1 cm, which is less than 0.4 inches. Two such frogs can sit together on an average person's thumbnail without touching each other. But which actual species wins the designation for world's smallest is difficult to determine.

Included among the world's midget frogs is *Eleutherodactylus iberia* of Cuba, a dark brown frog with a yellow stripe down each side and with no common name. Equally minuscule is *Brachycephalus didactylus*, a poorly known species from Brazil in the family Brachycephalidae. Some authorities consider *B. didactylus* and *E. iberia* to be smaller than any other terrestrial vertebrates in the world. Another minute species that reaches a length of

10–11 mm, which is still less than a half inch, is *Stumpffia tridactyla* (family Microhylidae) from Madagascar. *Sechellophryne gardineri*, a member of the family Sooglossidae, seldom grows above 11 millimeters, but a maximum size for the species of 13 millimeters (about a half inch) was reported from the Seychelles Islands by Ronald Nussbaum of the University of Michigan and Sheng-Hai Wu of National Chung-Hsing University, Taiwan. The larger size known for *S. gardineri* may simply be because the investigators measured more specimens and eventually found a "large" one. The same might happen with any of the other three species. The uncommonness of all four of these species can be noted because none have common names and are referred to by their scientific names only.

The species with the largest tadpole is probably the paradox frog (*Pseudis paradoxa*, family Hylidae) of South America. Although the adult frogs are no more than 3 inches in length, the tadpoles reach more than 8½ inches.

Why do so many frogs have long legs?

Frogs have comparatively longer legs, particularly hind legs, than any other group of vertebrate animals. Some species of frogs have much longer legs than others, but the hind limbs are longer than the front limbs in all species of frogs. Toads generally have shorter legs in proportion to their body than the true frogs or the treefrogs, but even toads have relatively longer hind legs than their front legs or than the hind legs of most other animal species.

Strong, elongated hind legs allow frogs to jump long distances relative to their body length. Jumping, which aids in rapid transportation, is especially effective for escaping from predators, thus offering a significant advantage to an otherwise vulnerable prey species. The single adaptation of jumping, or saltatory (proceeding by leaps) locomotion, is considered by many amphibian biologists to be responsible for the worldwide success, that is, high biodiversity, of this group of amphibians.

The powerful hind legs of some frogs are used not only for jumping but also for swimming. A large bullfrog can jump several times its body length from a riverbank into the water, and its strong legs and large webbed feet can then be used to propel the frog rapidly underwater. The long legs of treefrogs are used for reaching out to grasp the next leaf or tree branch.

Are frogs slimy?

Most frogs have moist skin, and frogs closely associated with aquatic habitats typically have slimier skin than those found on land in drier areas.

Tadpoles of the river frog (*Rana heckscheri*) are the largest of North American frogs. Courtesy John D. Willson

Some frogs produce mucus that makes them difficult to hold, which allows them to escape from some predators. All tadpoles are slippery.

Frogs have lungs, but many species also breathe by oxygen that diffuses from the air across the skin and into their bloodstream. For this type of respiration to occur, the skin must remain moist. In general, toads have drier skin than other frogs, but there are a number of exceptions. Many frogs and toads produce toxins in the skin that make them unpalatable to some predators. This toxin is mixed in as part of the "slime" on the back and legs of some poisonous frogs.

Are some frogs poisonous?

Many, perhaps most, frogs have at least some toxins produced by glands in their skin, and poison can be mild or highly potent, depending on the species. No frogs inject venom through fangs or stingers. Hence, some may be poisonous, but none are venomous.

The toxins produced by frogs' skin glands unquestionably serve as a defense against some predators by making them unpalatable. Toxic-producing glands all over the body and legs in some species provide protection no matter where a predator grabs the frog. Toads have a bumpy surface, and some bumps are toxin-producing glands. The toxic substances probably provide an additional service to the frog in some species by inhibiting parasites from attaching to the frog's skin and prevent any fungus or bacteria from colonizing.

A pair of parotoid glands are evident atop the head in some frog species, such as White's treefrog (*Litoria caerulea*) of Australia and toads in the family Bufonidae. These large glands produce toxic secretions that can sometimes be seen as a milky liquid if the gland is squeezed. Secretions from the parotoid glands of some species, such as the marine toad (*Bufo marinus*), can be extremely toxic and even lethal for some animals that bite them. The poisonous substance is called bufotoxin. Its effect on some potential predators is evident when a dog bites a common garden toad and then begins to foam at the mouth.

The toxic skin secretions of some frogs, such as common gray treefrogs (*Hyla versicolor*) or the Cuban treefrog (*Osteopilus septentrionalis*), which has been introduced into Florida, can cause extreme discomfort if secretions reach the membranes of the eyes or nose. A few South American species of frogs, known as the poison dart frogs in the family Dendrobatidae, produce alkaloid toxins on their skin that are among the deadliest poisons known. People have been known to go into a coma-like state and almost die after picking up one of these frogs if the toxins enter even a minor cut on their hand. If sufficient amounts of these toxins enter the bloodstream, they can be lethal to humans.

The most poisonous frog, and possibly the most poisonous terrestrial vertebrate, in the world is the golden poison frog (*Phyllobates terribilis*) of Colombia. Among the toxins produced by the species is batrachotoxin, a powerful alkaloid that can kill large mammals, including humans. When native people of the rain forests on Colombia's Pacific Coast rub the skin of golden poison frogs on their blowgun darts, they produce a lethal weapon that can kill an animal almost instantly upon penetration.

The color patterns of poison dart frogs and other species of Dendrobatidae include brightly colored blues, reds, and yellows, which are presumably warning colors to other animals that might try to eat them. Consequently, many brightly colored toxic species come out in the daytime, instead of at night like most other frogs, because their toxicity makes the risk of predation low. Poison dart frogs are presumed to acquire their toxins by eating certain kinds of invertebrates that produce batrachotoxins and other alkaloids.

Poison dart frogs prey on a particular group of beetles in the family Melyridae, several species of ants, and a species of millipede. The frogs apparently are able to sequester toxins from these insects without harming themselves. The level of toxicity varies considerably among species within the family Dendrobatidae and among other species of frogs worldwide. Interestingly, when poison dart frogs are kept in captivity and fed diets of crickets or mealworms, they eventually lose their skin toxicity. However, they retain their brilliant color patterns.

The eggs and tadpoles of some species of frogs and toads have been found to be unpalatable to some predators, such as fish and salamanders,

Frogs: The Animal Answer Guide

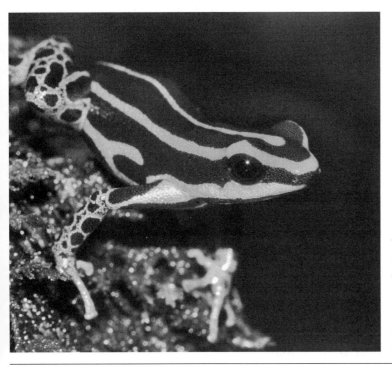

The Amazonian poison frog (*Dendrobates ventrimaculatus*) belongs to the family Dendrobatidae in which some species produce alkaloid toxins on their skin that are among the deadliest poisons known.

Courtesy Chris Gillette

but whether poisonous substances are involved is undetermined for most species that have been examined. The tadpoles of the golden poison frog are not poisonous, but juveniles become poisonous soon after metamorphosis when they assume a terrestrial diet.

Why do toads have so many bumps?

Toads are found mostly on land but breed in water. Their skin is much drier than the skin of other frogs, and the back and upper part of the legs of some species are covered with bumps that resemble warts, some of which are actually poison glands. Bumps are formed of keratin, the same protein found in human hair and fingernails, which makes toads less permeable to water, thus reducing water loss and allowing them to inhabit drier habitats. The pair of large bumps on the back of the head are called *parotoid glands*. These glands produce toxins exuded when the toad is bitten on the head, therefore discouraging many would-be predators. Although the pattern and arrangement of bumps on the backs of toads are likely of no advantage to the toad, often amphibian biologists use them to distinguish between species.

The numerous bumps with wart-like appearance on the back, head, and sides of a Great Plains toad (*Bufo cognatus*) aid in avoiding dehydration. Courtesy John D. Willson

Why is a frog's metabolism important?

Metabolism is the use of energy for life functions and is influenced by the body temperature of the animal. Frogs are ectothermic (cold-blooded) animals, so their body temperature and thus metabolism reflect the surrounding environment. Most endothermic (warm-blooded) animals, namely, mammals and birds, maintain higher and more stable temperatures year round and consequently have higher, more stable metabolic rates than ectotherms. Consequently, birds and mammals use much more energy during a year than ectothermic animals such as frogs use. Water, air, or soil temperatures surrounding a frog are major determinants of its body temperature, and its metabolism increases as its internal temperature rises. Thus, the body temperature of a Temperate Zone frog, such as a bullfrog (*Rana catesbeiana*), may vary during a year from near freezing in winter, resulting in a very low metabolism, to above 80 degrees Fahrenheit in summer, depending on environmental temperatures. When an animal's metabolism is high, it must eat frequently to acquire enough energy, whereas an animal with a low metabolism can go long periods without eating. A frog underwater or underground during cold weather also requires less oxygen. Frogs' low-energy lifestyle is efficient for energy conservation and allows frogs to allocate much of the energy acquired from food to growth and reproduction, rather than using energy to maintain a warm body temperature, like birds and mammals. A lowered metabolism is especially important during hibernation and aestivation. Some frogs, such as spadefoot toads (*Scaphiopus holbrookii*), can remain dormant underground at low temperatures for months or even years without eating by lowering their metabolic rates.

Do frogs have teeth?

Most frogs have small teeth, primarily maxillary (upper jaw) and vomerine (in the roof of the mouth) teeth. Only one species of frog, Gunther's marsupial frog (*Gastrotheca guentheri*) of Ecuador and Colombia, has teeth in the lower jaw. Species in the large family of toads, the Bufonidae, lack teeth. Frog teeth are small, mostly cone shaped, and are used to grasp their prey. Frogs do not chew their food but instead swallow their prey whole. They use their tongue, forelimbs, and even their eyes (by pushing them backward) to force captured prey backward into the mouth and down the throat. Toothless toads are evidence that teeth play a minor role in eating and are unnecessary. In general, frogs with teeth do not bite for defense but only for handling prey.

The African bullfrog (*Pyxicephalus adspersus*; family Ranidae) has what appear to be large, fang-like teeth: two large, bony spines on each side of the lower jaw separated by a smaller spine. These project upward from the lower jaw, and although they are odontoids, not true teeth but formed from bone, they function in the same manner. Although their teeth are not as large as those found in the African bullfrog, lower jaw odontoids can be found in the Sumaco horned treefrog (*Hemiphractus proboscideus*; family Hemiphractidae) of South America, in the Solomon Island leaf frog (*Ceratobatrachus guentheri*; family Ranidae), and in the tusked frog (*Adelotus brevis*; family Myobatrachidae) of Australia.

Most aquatic tadpoles have teeth-like horny plates in rows across the mouth in both the upper and lower jaws. The teeth are made of keratin and are often shaped for effectively scraping algae from rocks or other surfaces.

Do frogs sleep?

All frogs presumably sleep, and most species sleep primarily during daytime since most frogs are nocturnal. A sleeping frog closes its eyes. Many frogs and toads remain inactive belowground or in other hiding places for long periods in which they presumably rest in a sleep-like state. Hence, many frogs may actually sleep for well over half their lives. However, cold-induced hibernation differs from normal sleep because it is a long-term lowering of metabolic processes as a result of a lowered body temperature, so that the frog actually enters into a seemingly lifeless, unconscious state known as torpor from which it cannot be readily aroused.

The casque-headed frogs of the South American genus *Hemiphractus* have bony projections in the lower jaw known as odontoids. Although they are formed from bone and are not true teeth, they function in the same manner. Courtesy Dante Fenolio

Do frogs have ears and can they hear?

Nearly all frogs have ears, but frog ears are not external flaps of cartilage and skin like humans and most mammals have. The eardrum, or tympanum (which means drum), of a frog is visible on the outside of the head behind the eye. The tympanum is the same structure as our eardrum. In some frogs, such as bullfrogs and green frogs, the male's tympanum is much larger than the female's. Frogs hear extremely well, and hearing is an important part of the mating process in many species, as males call to indicate their location to females for breeding or to warn other males not to invade their territory.

Some frogs that live in the vicinity of waterfalls have higher-pitched calls and special hearing strategies to deal with the loud roar of the water. The hole-in-the-head frog (*Huia cavitympanum*; family Ranidae) of Borneo is the only frog known to produce a purely ultrasonic call, which the human ear cannot detect. The concave-eared torrent frog (*Odorrana tormota*; family Ranidae), which lives near noisy waterfalls in China, produces some ultrasonic sounds while making other calls inaudible to humans. The torrent frog's ability to adjust its hearing intentionally to select for different frequencies produced by calling males is believed to be a rare trait among vertebrates. Both frogs have unusual concave ear openings. Even the goliath frog has a much higher-pitched voice than would be expected of such a large species, presumably because it inhabits areas around large, noisy waterfalls. The Puerto Rican coqui (*Eleutherodactylus coqui*; family Eleutherodactylidae) has a specialized mechanism of hearing not only through the tympani but also through its body wall, which vibrates in response to the

The brightly colored Panamanian golden frog (*Atelopus zeteki*) is unusual in having no tympanum or middle ear cavity. Courtesy Dante Fenolio

males' calls. Receiving sounds from two different parts of the body allows for more accurate determination of the call's location.

Charlotte Steelman at Davidson College in North Carolina conducted a study to determine how man-made noise, such as airplanes, affected calling activity of five species of frogs. She found that calling ceased or was greatly reduced when planes flew over frog breeding sites and that most calling for all species occurred when there was no airplane noise. As a plane approached and the intensity of noise levels increased, fewer frogs called, and most calling activity for all anuran species occurred in the absence of airplane noise. Her findings suggest that a male frog is able to assess that the female is less likely to hear his calling during high-noise periods and that the benefits of calling are not worth the energy expended to make the calls.

How far can frogs jump?

Many frogs can jump at least 30 times their body length, and some smaller species of treefrogs can probably jump more than 50 times their body length. This would be equivalent to a human jumping the length of a football field without a running start (as if a running start really matters).

Although the exact record is disputed, a couple of larger species within the family Ranidae, which includes bullfrogs and leopard frogs, have been reported to jump more than 30 feet.

Because they actually glide through the air, some frogs in the genus *Rhacophorus*, the so-called flying or gliding frogs in Asia, can go the longest distances, depending on the height from which they jump. These frogs have extensive webbed toes that they extend and use as parachutes to slow their fall and glide to the next tree or to the ground. Because jumping long distances is important to the survival of many species of frogs for predator avoidance, the skeleton of some species is modified to absorb shock when they land, although the force of landing for small frogs is of little consequence. Although many frogs can jump long distances, some, such as the narrowmouth toads of the southeast or the Mexican burrowing toads, can only hop a few inches at most.

Can frogs climb?

Frogs of many families, including Hylidae, Rhacophoridae, and Centrolenidae, are excellent climbers, and some species seldom descend to the ground. Expanded toepads are found on many species that stick to leaves and tree trunks to aid in climbing, allowing these frogs to walk straight up a tree trunk or a plate-glass window. There is even a species of toad (*Bufo coniferus*; family Bufonidae) in Central America that has long fingers and is a good climber. A general rule is that frogs that climb are more slender bodied and smaller than those that do not climb. Climbing frogs, especially treefrogs, are obvious on man-made structures, such as windows, at night to which they stick with their toepads.

Can all frogs swim?

Most frogs swim, and some swim well. Some species, such as the African clawed frog (*Xenopus laevis*) and the Surinam toad (*Pipa pipa*) in the family Pipidae, are exclusively aquatic, spending their entire lives in the water. Male clawed frogs even call from underwater when mating. Most frogs have some webbing on their hind feet, and some have webbing on both the front and back feet to help with swimming. The extent of the webbing usually indicates a frog species' aquatic ability. Even the most terrestrial toads have some webbing on their back feet and can swim well. Some species of Indochinese cascade frogs, such as *Rana vitrea* from Laos, that live in fast-flowing streams have expanded fingertips that help them gain purchase on rocks in moving water. Many species use swimming as a

Even treefrogs that spend most of their life away from water, such as this barking treefrog (*Hyla gratiosa*), have webbed feet for swimming in wetlands during the breeding season. Courtesy Thomas Luhring

defense mechanism, wherein they swim to the bottom and bury themselves in debris to escape predators.

Can frogs breathe under water?

Except for a single species (the Kalimantan jungle toad [*Barbourula kalimantanensis*; family Bombinatoridae] of Borneo) all frogs have lungs and breathe air, but most can also breathe when they are under water as oxygen diffuses into the blood through the frog's skin. The most efficient oxygen exchange occurs through the skin at cold temperatures. Aquatic tadpoles breathe through gills under water and through their skin.

Some frogs have special and unusual adaptations to allow gas exchange across their skin. For example, the Titicaca water frog (*Telmatobius culeus*; family Ceratophryidae) is an exclusively aquatic species that lives at high elevations (above 12,000 feet) in Lake Titicaca in South America. The species stays permanently in the water and does not breathe air but has extensive skin folds that allow it to have greater surface contact with surrounding water and higher oxygen diffusion. The hairy frog (*Trichobatrachus robustus*; family Arthroleptidae) of Central Africa is noted for long, thin projections of its skin that grow from the sides and rear legs and gives the frog a "hairy" appearance. The species breeds in water, and the males, which remain with the eggs, use hair-like skin projections to increase their surface area and to allow for greater oxygen diffusion.

Form and Function

25

The Titicaca water frog (*Telmatobius culeus*) is an exclusively aquatic species that lives at elevations above 12,000 feet in Lake Titicaca in South America. Courtesy Dante Fenolio

The vocal sac of a calling male southern toad (*Bufo terrestris*) creates a resonant chamber that allows the trill to be heard over long distances during the mating season.
Courtesy Chris Gillette

What is the sac under the frog's throat?

The sac under a frog's throat is referred to as a *vocal sac*, which males use for calling during the mating season. The sac typically appears as a flattened patch but becomes a resonant chamber that increases the volume of the call when it expands. Because of the concentration of melanin when the vocal sac contracts, the area is often darker in adult males during the breed-

Frogs: The Animal Answer Guide

ing season than in females or juvenile males. The vocal sac can thus be used to distinguish between the sexes of some species.

Although the single large vocal sac beneath the throat characterizes many species of frogs and toads, some species (e.g., leopard frogs, *Rana pipiens*; family Ranidae) have paired vocal sacs that protrude from the side of the throat. Vocal sac size relative to the size of the frog varies among species. Some sacs can inflate to a size larger than the frog. Vocal sacs of some species take on specialty roles. The vocal sac of the male Darwin's frog (*Rhinoderma darwinii*; family Cycloramphidae) from Chile is used as an incubation chamber for developing eggs. Species in the genus *Heleioporus* (family Myobatrachidae) of Western Australia do not have vocal sacs but can enlarge the inside of the mouth to create a resonance chamber. Males call from burrows that further amplify the sound. Several species in the genus *Neobatrachus*, which is widespread in Australia, also have no vocal sacs.

Chapter 3

Frog Colors

Why are frogs colored the way they are?

Few other animal groups display the variety of colors found among frogs and toads. Frogs come in greens, reds, browns, yellow, and sometimes blue. The diversity of colors found among frog species, and even within a single frog species, is remarkable. In nearly all cases, the colors and patterns frogs exhibit have some adaptive value for the animal. In most cases, frogs' colors protect them from predation. Patterns and colors can camouflage, warn, and deceive other animals or even startle potential predators.

Perhaps one primary reason for frogs' coloration is to conceal. Many species of frogs have patterns and colors that camouflage to reduce the chance of detection by potential predators or, in some cases, so that they can be concealed from potential prey. Toads are normally mottled brown or gray because they are nearly always found on the ground where grays and browns dominate the color spectrum. Treefrogs are often green to blend in with vegetation. Some treefrogs, such as Cope's gray treefrogs (*Hyla chrysoscelis*), are mottled gray and are nearly invisible when resting on tree bark. Some frogs have elaborate camouflage coloration aided by skin texture. The moss frog (*Theloderma corticale*) from Vietnam is one such animal. It is black and green to look like the mossy trees in which it lives, and fleshy projections on its skin resemble the fibrousness of mosses, which help it to nearly disappear into a rain forest background. Several species of frogs remarkably resemble fallen leaves and blend in well in the forest leaf litter where they live.

Instead of blending into their environment, some brightly colored frogs quite visibly stand out. Such species include many of the poison dart frogs

Because of the lack of skin pigment from a genetic condition known as leucism, this eastern narrowmouth toad (*Gastrophryne carolinensis*) would likely not survive in the wild.

Courtesy John D. Willson

(family Dendrobatidae) from Central and South America. Because the skin of these tiny frogs is highly toxic to nearly any animal that eats them, their coloration warns potential predators to stay away. Other brightly colored species include some of the mantellas from Madagascar. Like the poison dart frogs, these frogs are small, brightly colored, and toxic.

Some frogs that are not toxic or dangerous mimic toxic animals, including other toxic frogs. In general, such a phenomenon is known as *Batesian mimicry*. The sanguine poison dart frog (*Allobates zaparo*) of Ecuador and Peru is actually not toxic but looks similar to other toxic dart frogs, which in this type of mimicry are referred to as *models*. Presumably, such coloration protects the frog from predation by animals aware of the toxic model.

A few frogs have patterns or colors that are actually designed to deceive other animals, primarily potential predators. The false-eyed frog (*Eupemphix nattereri*) of Brazil has large dark spots on the posterior part of its body. When threatened, this little frog will raise its rear end toward the predator so that the dark spots appear to be two large eyes. Apparently, this dissuades the predator from attacking the little frog. If such a display is unsuccessful in preventing an attack, glands in the frog's groin area produce a noxious secretion that will generally discourage a predator from ingesting it.

Other species have bright colors or patterns that are hidden when the animal is at rest but are visible when the animal moves. Such coloration is generally referred to as *flash coloration*. Common gray treefrogs (*Hyla versicolor*) have orange or yellow on their thighs that is only visible when their legs are extended. It is thought that the sudden appearance of this bright color made visible to a predator by the frog's movement will startle the predator, thus providing time for the frog to escape. Also, the predator may continue searching for the frog but is looking for bright yellow instead of the now-camouflaged treefrog as it sits on the tree trunk. Other treefrogs have similar coloration, such as the red-eyed treefrog (*Agalychnis callidryas*) of Central America, which has blue and white banding along its sides only visible when it extends its legs. Fire-bellied toads (genus *Bombina*) of Europe and Asia have camouflaged backs but bright red and black bellies. When threatened, they exhibit a posture known as the *unken reflex* in which they arch their back and raise their legs to display the bright color on their underside. Apparently, the underside coloration warns predators that fire-bellied toads are foul tasting.

Many animals and nearly all frogs are light underneath and darker above. Such a phenomenon is known as *countershading*, or *Thayer's Law*. Basically, such coloration makes the shadows of the animal blend in better with its surroundings. If a frog is floating on the water's surface, countershading causes the frog to be less visible from below because it blends in better with the light sky. Likewise, it would be less visible from above because it blends in better with the darker water, or bottom, of the water body. Some scientists also think that the darker coloration of animals' backs, including frogs, helps protect them from dangerous effects of ultraviolet radiation.

In some species of frogs, coloration may depend on their environment. For example, research by Janalee Caldwell showed that the tadpole of the northern cricket frog (*Acris crepitans*) exhibits a polymorphism in the color of its tail in response to different predation pressure. Tadpoles from ponds have black tails that apparently direct dragonfly larvae attacks to the tail rather than the head. Whereas tadpoles from streams and lakes, where dragonfly larvae are less common but fish are major predators, are plain.

Although most frogs and toads have colors that help them to survive, on rare occasions, some individuals have aberrant coloration that, in some cases, may decrease the likelihood the frog will survive. In nearly all cases, such coloration is the result of a recessive gene likely selected against strongly. Albino individuals, those lacking dark pigment, have been found in many species. Some normally green frogs will appear abnormally blue because they lack pigment.

What causes the different skin colors of frogs?

Pigment cells within the skin cause frogs' skin colors. Various types of pigment cells, called *chromatophores*, can be present, and how they are arranged results in a frog's color. Many frogs are green. Green color in frogs is a result of three different types of chromatophores and their arrangement. Within the skin of the frog, the deepest chromatophores contain dark pigment known as *melanin* and are known as *melanophores*. Above the melanophore layer is a layer of cells known as *iridiphores*. Iridiphores do not contain pigment but instead contain purine crystals. When light hits these cells, blue light is reflected back. Blue light passes through the outermost layer of chromatophores, known as *xanthophores*, which contain yellow pigment, and the reflected light appears green to the observer. Because chromatophores can change shape, a frog can alter its color, depending on temperature, activity, and other factors. For example, when a frog expands it *melanophores*, it becomes darker because the melanin within these cells is distributed over a wider area.

What color are a frog's eyes?

There are two parts of a frog's eye that determine the colors we see. The pupil, which is the opening in the eye that allows light to pass to the retina, appears black. In some frogs, such as spadefoot toads, the pupil is vertical, and in others, such as true toads of the genus *Bufo*, the pupil is horizontal. The iris surrounds the pupil and can be colored and patterned in numerous ways. In humans, the iris can be brown, gray, blue, or sometimes green. However, frogs display a remarkable range of colors from bright red in the red-eyed treefrog (*Agalychnis callidryas*) to brown or gold. Some species have blue irises and some have a golden iris with patterns of black. In some frogs, such as poison dart frogs, the iris is entirely black. Many frogs have bands of coloration running through their eyes that are continuations of dark bands on the sides of their heads. Such coloration may conceal the eye and make the animal more difficult to detect. Most tadpoles have solid black eyes, but some may be brown or golden in color. River frog tadpoles (*Rana heckscheri*) have bright red eyes that change to more typical golden brown when they become adult frogs. The adaptive value of having red eyes as a tadpole is unknown.

Do a frog's colors change as they grow?

When a tadpole undergoes metamorphosis and becomes an adult frog, remarkable changes occur. In nearly all species, the adult frog's color and pattern are different from the tadpole. As a newly transformed frog grows to become a sexually mature adult, it may undergo some changes in color and pattern, but generally, these are less drastic than those associated with metamorphosis. Most tadpoles are not as brightly colored as adults and are either dull brown or greenish brown. As tadpoles, closely related species are often difficult to distinguish, even by experts in the field. As metamor-

The treefrog *Agalychnis callidryas* of the American Tropics has large, bright red eyes. Courtesy John D. Willson

phosis progresses, the coloration begins to look more like the adult. Over time, both front and hind legs are present, and the species can be recognized by its adult color and pattern. Although the intensity of colors may change within short time frames, most frogs' colors rarely undergo substantial change once they are adults.

Do a frog's colors change under different conditions?

Although the colors and patterns of frogs are pretty well set as adults for most frogs, the intensity (i.e., darkness) of their skin can rapidly change. In general, varying environmental conditions and activities of frogs result in changes in their colors. Hormonal changes are likely the physiological mechanism responsible for color change. Some frogs, such as the common gray treefrog (*Hyla versicolor*) and barking treefrog (*H. gratiosa*), can change from brown or gray to bright green under certain conditions. In general, when most frogs are active, they tend to be lighter than when they are resting or dormant. For example, when a hibernating frog is uncovered, it is often dark brown or even nearly black. Some frogs change color to match their background, which provides protective camouflage. Lee Kats and Randall Van Dragt conducted a study in which they put spring peepers (*Pseudacris crucifer*) on white and black backgrounds and showed that the frogs would darken or lighten to better match the background on which they rested. When they were placed on white backgrounds with dark stripes, the frogs became neither totally dark nor light but instead exhibited a background color of intermediate intensity. Some frogs may become dark as they bask in the sun. Such color change likely allows them to effectively absorb solar radiation, thus increasing their body temperature. Although most frogs do not undergo as remarkable color changes as some other ani-

Frogs: The Animal Answer Guide

Although the reasons are not fully understood, individual frogs of many species are capable of changing color as with this common gray treefrog (*Hyla versicolor*). The same individual can vary from light green (*left*) to dark brown (*right*). Courtesy Aubrey Heupel

mals, such as lizards, squid, and octopi, the ability to change color based on background likely provides important protection from potential predators.

Do all individuals of a single frog species look the same?

Within a species, most frogs look relatively similar to one another. When frogs of the same species are not similar in either color or morphology, it is referred to as a *polychromatism*, or *polymorphism*. Some species are well known for their pronounced polymorphisms. At least 225 species within 35 genera and 11 families have been documented to have color or pattern polymorphisms. The genetics governing such polymorphisms is unknown in most cases, but simple Mendelian inheritance has been demonstrated to occur in a few species. At least 32 species of frogs and toads have sexually based polymorphism in which the female and males look different. Such species are referred to as *sexually dimorphic*. The famous golden toad (*Bufo periglenes*) of Costa Rica, which appears now to be extinct, gets its name from the solid golden yellow color of adult males. Females, however, do not look anything like males, and are a mixture of browns, grays, and greens, with many red spots. Why males and females differ in color is unknown but may be related to sex recognition. The ornate chorus frog (*Pseudacris ornata*) is found in the Coastal Plain of the southeastern United States and can occur in at least four color morphs (gray, brown, green, and red). Research by Gail Harkey and Ray Semlitsch found that temperature during development may play a role in determining the color phase of adult frogs.

In some species of frogs, variation may occur related to the geographic region in which they are found. For example, the green frog (*Rana clamitans*) is green in the northern part of its range and shiny brown in the southern part of its range, where it is often called the bronze frog. Some individuals of some species may vary quite a bit in color pattern. Southern leopard frogs (*Rana sphenocephala*), which get their names from the spots on their back, sometimes have few spots. Sometimes, green treefrogs (*Hyla cinerea*), which typically have a silvery white stripe along the side of their body, have no such stripe. Some individuals may have numerous tiny orange or gold spots on the back, whereas others do not. Although generally not as colorful as many other species, toads of the genus *Bufo* can exhibit substantial intraspecific variation in color pattern. The American toad (*Bufo americanus*) is generally gray or brown but can sometimes be brick red.

Frogs: The Animal Answer Guide

Chapter 4

Frog Behavior

Why do frogs make noise?

Frogs make a lot of noise. The noise a large chorus of frogs produce can be deafening. Why do they make so much noise? Frogs are trying to secure a mate and calling, sometimes very loudly, is the main way they do this. In nearly all species, primarily the male calls to attract a mate. Such a call, which scientists refer to as an *advertisement call*, cannot only attract females but also, in many species, can warn other males that they are encroaching on the calling frog's territory. The advertisement call is unique to each species—that is, a frog can nearly always be identified by its advertisement call. Male frogs gather near a suitable breeding site to call at a particular time of year known as the species' breeding season. Female frogs can detect slight variations in males' calls and are more attracted to males with calling characteristics that denote high reproductive fitness. Calling in most species occurs primarily at night but may occur during the day, especially for those that breed during colder times of the year.

Although advertisement calls are best known and are the type of call most easily recognized by humans, most frogs and toads produce other types of calls for reasons other than advertising to attract mates. For example, many species produce a release call if grabbed by a male frog attempting to mate with an individual that is not receptive to mating (e.g., other males or females that have already laid their eggs). The release call informs the mating male that the union is unacceptable. Some frogs, such as bullfrogs, have specific territorial calls. If approached too closely by another male, bullfrogs will produce a quick "phoot" call to preserve their territory. Many frogs, especially frogs in the family Ranidae, will emit a

Most noises made by frogs are by males that call during the breeding season. However, some species will emit a scream when being attacked by a predator, such as this Cuban treefrog (*Osteopilus septentrionalis*) being eaten by a garter snake (*Thamnophis sirtalis*). Courtesy Chris Gillette

loud "squeak" when startled. Other species, such as squirrel treefrogs (*Hyla squirella*), produce "rain calls" during the afternoon when summer rainstorms are approaching. In some species, such as the midwife toad (*Alytes obstetricans*) of Europe, the female will produce a call in response to the male, indicating she is willing to mate. The male then changes its advertisement call as a response to the female.

A remarkable call emitted by many species of anurans is the distress call. The distress call of some species can be loud and is produced when the animal is threatened or grasped by a predator. Distress calls are made with an open mouth. The smoky jungle frog (*Leptodactylus pentadactylus*) is found in the New World Tropics and is well known for its large size and loud distress call that sounds like a traumatized cat. The bullfrog (*Rana catesbeiana*) also produces a loud distress call, most often when it is swallowed legs first by a watersnake. The loud scream is similar to the sound a rabbit emits when it is attacked. Many scientists think frogs' distress calls attract predators that may attack the animal trying to eat them or at least disrupt the event and allow them time to escape.

How do frogs make sounds?

Frogs produce sounds similar to how humans vocalize; that is, air is passed over vocal chords that vibrate, producing sounds. Frogs have some anatomical modifications that help to increase the volume of their call, thus making them discernable to potential mates for long distances. The primary structure that increases the volume of frog calls is the vocal sac (or

Frogs: The Animal Answer Guide

Most frogs, such as this common gray treefrog (*Hyla versicolor*), have a single vocal sac below their chin that expands when inflated with air and acts as a resonance chamber to amplify the sound they produce.

Courtesy Thomas Luhring

sacs), which acts as a resonance chamber to amplify the sound they produce. When calling, the vocal sac is inflated with air and expands greatly. Many frogs have a single vocal sac below their chin, but some species, such as northern leopard frogs (*Rana pipiens*) and the edible frog (*Rana esculenta*) of Europe, have paired vocal sacs on either side of their head.

Frogs can call from various locations, including from the ground, on vegetation, or while floating in the water. Calling, however, can attract not only mates but also potential predators. Thus, frogs frequently stop calling if they are disturbed, and it may take a while for them to resume calling. Sometimes, however, the urge to call is so great that some species will call regardless of how much they are disturbed. Some frogs decrease their risk of predation by calling from hiding places, such as within areas of thick grass. Some decrease their chances of being heard by predators by calling under water. Typically, such frogs have low-frequency calls that transmit best under water. Frogs known to call under water include gopher frogs (*Rana capito*), spotted frogs (*Rana pretiosa*), and African clawed frogs (*Xenopus laevis*).

Frogs may communicate in many ways other than typical calling. One of the most interesting is seismic communication. Research by Edwin Lewis and colleagues showed that Puerto Rican white-lipped frogs (*Leptodactylus albilabris*) thump the ground with their feet and that other members of the species within close range will respond to the signals transmitted through the ground. A special saccule in the frog's ear receives the seismic signals.

Calling by male anurans results in other costs in addition to the risk of predation. Most notably, calling can be very energetically expensive. Adjusted for their size, many species of frogs produce some of the loudest

Some frogs, such as the carpenter frog (*Rana virgatipes*) have paired vocal sacs on either side of their head. Courtesy Thomas Luhring

calls of any animal. Sound production appears to be the most energetically expensive activity in which many species engage. Researchers have shown that calling frogs expend 10 times the energy of resting frogs.

Can frogs change their calls?

Many species of frogs change their calls because the costs or benefits of calling can be high or low, depending on the situation. On the one hand, the benefits frogs receive from calling are substantial; they attract potential mates. On the other hand, the costs associated with calling include high energetic demand and an increased risk of predation because the frog may reveal its location. Because of this predation risk, some frogs have developed the ability to alter their calls so that they still receive reproductive benefits but lessen their chance of becoming a meal for another animal. Classic work on this phenomenon was conducted by Michael Ryan on a little frog from Central America called the tungara frog (*Engystomops pustulosus*).

Like many other frogs, the small, bumpy-skinned tungara frog calls from small puddles and wetlands to attract mates. Its call can be simple or more complex, consisting of several different sounds, including "whines" and "chucks." The more complex the call, the more attractive it is to females. However, the male tungara frog must balance the ability to attract females with the potential for being detected by predators. Major predators of these frogs are frog-eating bats (*Trachops cirrhosus*), which locate the frogs based on their call. In addition, blood-sucking flies (*Corethrella* spp.) can locate and attack calling frogs. It is much easier for both bats and flies to find and attack frogs when they are producing more complex calls. Male

Frogs: The Animal Answer Guide

tungara frogs are much more likely to produce complex calls when calling in a group of other males than when calling alone. If they are calling alone, they have little competition for mates and, thus, can still attract females with rather simple calls. If they are calling in a chorus of other males, the likelihood that any individual frog will be attacked by a predator is low, but competition for females is much higher. The conundrum for male tungara frogs is to effectively balance between reproductive success and the potential to be eaten and is a classic example of the potential conflicts that can arise between natural selection and sexual selection in many animal species.

Environmental factors can affect the advertisement call male frogs and toads produce. In general, temperature slows the call notes and may affect the frequency (or pitch) of the call. One author of this book (MED) showed that southwestern toads (*Bufo microscaphus*) will reduce calling during rainy periods. This reduction in calling is likely influenced by the noise rain generates, which reduces the likelihood that a female will hear the males' calls.

Some frogs increase their volume even more by calling from particular locations. Björn Lardner and his colleagues have shown that male Bornean tree-hole frogs (*Metaphrynella sundana*) produce advertisement calls from tree cavities partially filled with water. Frogs tune the pitch of their call to match the size of the hole so that their call is most attractive to females. When a female approaches and selects a male, she lays eggs in the water at the bottom of the tree hole where the male fertilizes them.

Do all frogs make sounds?

Although frogs and toads are well known for their vocal abilities, a few frogs do not produce auditory calls. The Pacific tailed frog (*Ascaphus truei*) lives in mountain streams in the Pacific Northwest and does not vocalize. Males have no vocal sacs and apparently use visual cues to attract females. Several species of the primitive group of New Zealand frogs (family Leiopelmatidae) lack a vocal apparatus, although they produce small clicks during encounters with females. The South American Surinam toad (*Pipa pipa*) also has no vocal apparatus but instead uses its hyoid bone, found in the throat region, to make a snapping sound that attracts females. Of course, female frog species are for the most part silent, and no tadpole is known to produce any sort of vocalization.

Are frogs social?

Frogs and toads are social animals in that they congregate and interact with other members of their species. During breeding season, many individuals of most species of frogs and toads congregate in breeding aggregations. The interactions among individuals in these aggregations are complex and include males vying for the attention of females while often attempting to dissuade other males from coming too close. For some species, these social interactions may last for months and for other species perhaps only one night out of an entire year. Social interactions are often mediated through complex communication systems that involve advertisement calls to attract females, territorial calls to dissuade other males from encroaching, and release calls to notify potential suitors that a frog is not receptive to mating. Some may even include visual displays, such as waving their feet or pulsating their throat rapidly, to communicate with others of their species.

An interesting social interaction related to reproduction is the strategy some individual males of some frog species use in which they do not call but position themselves close to a calling male. Such males are known as *satellite males* and will often grab a female as they approach a calling male. This strategy allows males employing it to gain benefits of calling without incurring the energetic costs.

Frogs and toads may congregate for reasons other than breeding. Toads will often congregate around streetlights at night to prey on insects attracted to the light. Likewise, it is common to see many treefrogs stuck to windows at night where they have an abundant supply of insect food. Tadpoles of many species are social and congregate for protection. Toad tadpoles will often stay in dense schools, and tadpoles of the river frog (*Rana heckscheri*) will sometimes move together in schools large enough to be mistaken for an alligator moving under the water.

Do frogs fight?

We seldom think of frogs as fighting one another, but some species of frog actually use physical combat in social interactions. The large African bullfrog (*Pyxicephalus adspersus*) is well known for violent fights among males at breeding time. This species lives in dry grassland areas, where suitable breeding sites can be limited. Consequently, males fight off other males by grasping them with their jaws and flipping them over. The bony, formidable tooth-like projections in front of their lower jaw are used to grasp other frogs. They will also defend their tadpoles against potential predators and

A female green treefrog (*Hyla ci-nerea*) has responded to a calling male but been grabbed by a satellite male that will try to mate with her.
Courtesy Trip Lamb

The Senegal running frog (*Kassina senegalensis*) is a member of the family Hyperolidae and is found throughout much of sub-Saharan Africa. This species calls in large choruses and males are known to respond to each other's calls in attempts to vie for females. Courtesy Graham Alexander

The tadpoles of some frogs, such as the river frog (*Rana hecksheri*), are social in that they will stay together in large schools. Courtesy Robert T. Zappalorti

can bite hard enough to bloody the finger of a human. If threatened, they will puff themselves up and lunge open-mouthed at an intruder.

Combat among male frogs has also been documented in several species of arboreal glass frogs (family Centrolenidae). These male frogs will grasp other males belly to belly while holding on to vegetation. Presumably, the winner has a higher probability of mating success. Even some of the smallest frogs—poison dart frogs—commonly engage in physical combat. Males of the Ecuadorian poison dart frog (*Epipedobates tricolor*) and the strawberry poison frog (*Dendrobates pumilio*) are extremely territorial and will aggressively defend their breeding sites by chasing other males and by physical combat.

Frogs: The Animal Answer Guide

Males of this species, the African bullfrog (*Pyxicephalus adspersus*), use bony, tooth-like spines to grab and fight other males during the breeding season. In some cases, the fighting can result in the death of the loser. Courtesy Graham Alexander

Do any frogs fly or glide?

No frog can fly, at least not like a bird. However, several species of frogs have developed the ability to glide or parachute, and a few are quite proficient gliders. Many small treefrogs frequently jump from high trees to descend to the ground, sometimes falling from heights that would kill or injure a human. However, because of their small body size, falling from such heights does them no harm. By extending their limbs, they can slow the rate of their fall, further reducing the risk of injury. Some frog species have developed enlarged limbs and extensive webbed feet that enable them to glide. Some have extra flaps of skin on their elbows and ankles that slow their fall. These frogs use gliding to descend rapidly to the ground or to escape from predators. It is not surprising that gliding has evolved in several lineages of frogs. Many frogs already have some webbing between their toes, and thus, selection for more expansive webbing could result in the morphological features seen in these species.

How do some frogs stick to walls?

Many species of frogs have enlarged toepads that allow them to stick to vertical surfaces such as leaves, tree trunks, and glass. Toepads have evolved in many different families of frogs but are most well known in the treefrog families (e.g., Hylidae). Toepads are often wet with water and mucus from glands on the toes, and researchers have determined that capillary adhesion is responsible for the ability of frogs with toepads to stick to slick surfaces so effectively. Although such adhesion is likely important, Walter Federle and his colleagues used sophisticated techniques to show that the ability to attach to surfaces does not necessarily depend on water and mucus but

The flattened toes and extensive webbing between them allows Rabb's fringe-limbed treefrog (family Hylidae; *Ecnomiohyla rabborum*) of Panama to glide from the tops of high trees to the ground. Courtesy Dante Fenolio

is substantially enhanced by close contact and friction between the toepad skin and the substrate. Such high levels of friction result from the microscopic structure of toepad cells.

How smart are frogs?

Determining whether frogs are smart is difficult because humans perceive intelligence based on our perceptions of the world and the ability to learn. Because frogs and toads live in environments and are faced with challenges completely foreign to people, extrapolating human measures of intelligence on frogs may be unfair. Frogs respond to their environment by instinct, and learning likely plays a minor role in their behavior. However, there are some examples of what appears to be clearly learned behavior in frogs. Frogs and toads often congregate at night to eat insects attracted to lights. Such behavior is likely learned. Andrew Blaustein and Bruce Waldman showed that tadpoles of some species of frogs and toads likely learn to recognize their own siblings, although presumably this would have an innate component as well. Other researchers have shown that tadpoles are able to sense chemicals released by other tadpoles injured by predators and can learn to avoid predators based on such stimuli.

Do frogs play?

Play is generally defined as the process of occupying oneself in an activity that results in amusement or fun. Although animals such as dolphins and chimpanzees may exhibit numerous behaviors classified as play, there is no evidence that any species of frog or toad engages in playful activities.

Frogs: The Animal Answer Guide

Children and adults may play with recently captured or pet frogs, and we have engaged in such activities; however, in all likelihood, frogs simply tolerated our behavior and received little amusement as a result.

How do frogs defend themselves?

Most frogs and toads defend themselves by remaining undetected either through camouflage or simply by hiding out of sight beneath various objects or underground. If detected, many frogs will attempt to escape by jumping, and, of course, some can jump considerable distances. Some species may remain motionless to feign death. American bullfrogs, when captured, will often go completely limp in a person's hand. Whether this represents actual death-feigning behavior is equivocal. Many species of frogs will use bluffing to defend themselves. They will often fill their lungs with air to make themselves appear larger than they actually are and will sometimes raise their bodies off the ground. Many frogs also inflate themselves to prevent potential predators, especially snakes, from swallowing them. Some, such as African bullfrogs, will lunge toward an attacker with an open mouth.

Some frogs defend themselves using toxic chemicals. Many species of toads have toxic skin secretions released from glands on their necks called parotoid glands. These glands produce bufotoxins that can kill some animals that ingest them. Dogs have learned the hard way that biting a toad can sicken them. Of course, poison dart frogs are well known for their highly toxic skin secretions. Among poison dart frogs, toxicity varies; many species are only slightly toxic. The golden poison frog (*Phyllobates terribilis*) of Colombia is considered one of the most poisonous vertebrate animals in the world. Its alkaloid poison interferes with nerve transmission, and it can kill an animal that consumes it or receives it into its bloodstream through a cut or other wound.

Other species of frogs and toads use other mechanisms to protect themselves. Spadefoot toads will often emit a foul odor when handled. Toads in the genus *Bufo* will frequently empty the contents of their bladder when handled. This is presumably a defensive behavior but might just be a response to stress and have little adaptive value for the toad. The false-eyed frog (*Eupemphix nattereri*) of South America is a diminutive frog with two large spots on its hind end that resemble the eyes of a large animal. If threatened, the frog will raise its rear end to show its "eyes," which apparently threatens an intruder.

The rolling toad (*Oreophrynella nigra*) from Venezuela lives in montane habitats and, when threatened, tucks it legs under its body to form a ball. The "ball" then simply rolls away down the slope of the mountainside, escaping the potential predator. The frog is colored so that it looks like a rock and is often called the pebble toad.

Do any frogs bite to defend themselves?

Although most frogs do not bite to defend themselves, some species will bite if attacked by a potential predator. The African bullfrog and horned frogs (genus *Ceratrophrys*) are large and have proportionately larger mouths adapted for biting hard on prey. They will sometimes use their ability to bite to defend themselves against a predator. One species of marsupial frog, the Sumaco horned tree frog (*Hemiphractus proboscideus*), of South America will often gape at predators and may bite with "fangs" on its lower jaw, which are actually composed of bone.

Frogs: The Animal Answer Guide

Chapter 5

Frog Ecology

Which geographic regions have the most species of frogs?

With few exceptions, frogs and toads are found worldwide, and at least one species lives in nearly every region and frogs live on all but one continent. Most species live in the Tropics, and none occur in Antarctica, Greenland, or Iceland. Because they cannot tolerate salt water, no frogs are found in marine habitats except for a species that lives close to marine habitats. This is the crab-eating frog (*Fejervarya cancrivora*) of Malaysia that lives near and lays its eggs in brackish water. In Temperate Zone regions, the numbers of frog species are highest in warm, humid areas, such as the southeastern United States, and are lower in arid regions, at high altitudes, and at high latitudes where temperatures are colder.

What types of frogs live in rain forests?

Tropical rain forests have more species of frogs than any other habitat in the world, and many of them are arboreal. Among the most common families found in rain forests are the Hylidae, Leptodactylidae, and Centrolenidae (glass frogs) in the Western Hemisphere and Australia and the Rhacophoridae in Asia and Africa. Some rain forest frogs are burrowing species, such as narrowmouth toads (genus *Gastrophryne*; family Microhylidae) and species that live in the leaf litter. The Dendrobatidae (poison frogs) have approximately 180 species in the American Tropics. Toads (family Bufonidae) and true frogs (family Ranidae) occur in many tropical areas. Glass frogs and poison frogs are endemic to rain forests, but most

Some frogs, such as the Southeast Asian toad (*Bufo melanostictus*), have extensive geographic ranges. The species is found from eastern China through India to Sri Lanka.

Courtesy Cris Hagen

other groups have species found in the Temperate Zones. Other frog families found entirely or almost exclusively in rain forests are the Allophrynidae with one species, Ruthven's frog (*Allophryne ruthveni*), and Pseudidae in South America.

What types of frogs live in deserts?

Many frogs have adapted to living in dry areas and have different mechanisms for doing so. Some, such as the American spadefoot toads (genera *Spea* and *Scaphiopus*), live in arid or true desert conditions, stay underground for much of the year, and only emerge in large numbers during heavy rains for breeding. Thus, they survive the harsh conditions by coming to the surface only opportunistically and for relatively short periods. Spadefoots have been documented to remain underground for months or more than a year and possibly can do so for several years. Many frog species deal with drought or arid conditions by going underground and forming a cocoon that covers almost the entire body surface, including mouth, eyes, and cloaca. Only the nostril openings are not covered, allowing the animal to breathe while surrounded by the cocoon. The cocoon consists of an accumulation of multiple layers of single-cell-thick sheets of outer epidermal cells produced when the animal is exposed to dry conditions. Cocoons are common in many Australian desert species.

Some frogs that live in deserts have special characteristics such as the ability to store water. The water-holding frog (*Cyclorana platycephala*; fam-

Frogs: The Animal Answer Guide

The rufous-sided sticky frog (*Kalophrynus pleurostigma*) lives in the leaf litter of tropical lowland rain forests in Indonesia. Courtesy Cris Hagen

ily Hylidae) and the desert spadefoot toad (*Notaden nichollsi*; family Myobatrachidae) of Australia burrow as much as 3 feet beneath the ground during dry periods but are able to store enough water in their urinary bladder to increase their body weight by 50 percent. The aboriginal people of Australia have even been reported to capture water-holding frogs during periods of drought and squeeze them until they release water they can drink, similar to how a desert traveler might rely on a canteen. Toads in the family Bufonidae that live in the desert are able to combat dehydration by having a mostly impermeable skin so that water can be retained and not lost through the skin. These abilities allow these species to go for long periods without requiring water from external sources.

Another factor that can affect the success of frogs living in deserts is the frequency and limited number of available breeding sites, which can result in a competitive advantage of one species over another. Gage H. Dayton and Lee A. Fitzgerald studied four species of desert toads at Big Bend National Park, Texas, and found that Couch's spadefoot (*Scaphiopus couchii*) was dominant over three other species (western narrowmouth toad, *Gastrophryne olivacea*; Texas toad, *Bufo speciosus*; red-spotted toad, *Bufo punctatus*) during the tadpole stage. The investigators concluded that the competitive ability of spadefoots in ephemeral desert breeding sites usually excluded any other species. Some desert-dwelling African species of frogs in the genus *Breviceps* avoid high temperatures and dry conditions by laying their eggs in an underground nest or burrow. Metamorphosis of the tadpoles takes place inside the egg capsules, and the young emerge as fully formed froglets.

Why do so few frogs live in large lakes?

Frogs are usually absent from the open waters of large lakes because of predatory fish that eat the frogs' eggs, tadpoles, and small adult frogs. When frogs live in large lakes, they are often concentrated in shallow, heavily vegetated coves that provide hiding places for tadpoles. Also, the surface area of large, deep, clear lakes is much less productive than shallow wetlands, so less algae is available on which tadpoles feed. A few kinds of frogs, such as bullfrogs, are successful in large lakes because of their large body size as adults and as tadpoles. Some toads with toxic skin glands may be poisonous or at least distasteful to some fish as adults or possibly even as tadpoles and therefore may be able to breed successfully in open water along lake margins. Michael R. Crossland and Ross A. Alford reported that cane toads (*Bufo marinus*) introduced into Australia lay eggs that are poisonous to some animals. Although Australian predators would have had no previous exposure to cane toad eggs and hence would be poisoned after eating them, predators in their native habitats in tropical America would probably avoid the eggs if they were toxic. It is possible that eggs and tadpoles of other frogs are also toxic to some animals, but even these would probably live nearshore because of the low productivity in deeper parts of the lake.

Do any frogs live in salt water?

No frogs or tadpoles live in the ocean. The pressure of predation would presumably be extremely high on slow-moving tadpoles, but a more immediate physiological problem for most amphibians is their inability to deal with salt water. Because the concentration levels of salts in the blood and other tissues of frogs and tadpoles are less than that in salt water, water would diffuse rapidly across their skin, and they would dehydrate. At the same time, salts would diffuse into the frog's body until they reached toxic levels. One species (crab-eating frog, *Fejervarya cancrivora*) lives in brackish (partially salt) water in mangrove areas in Malaysia. Instead of excreting ammonia as a waste product like most frogs, the crab-eating frog retains urea in its body to make its ionic concentration higher, thus reducing the amount of water that leaves across its skin. Sharks use the same physiological strategy to avoid dehydration. Although they develop mostly in brackish waters, the tadpoles of the crab-eating frog are able to persist in full seawater and are able to regulate their body water and salt concentrations without retaining urea. Three other species, the green toad (*Bufo viridis*), the southern leopard frog (*Rana sphenocephala*), and the African clawed frog are known to inhabit brackish water at times.

Do frogs migrate?

Migration generally refers to movements of individual animals from one location to another, followed by their return to the former location. Migration usually occurs because one habitat is suitable for one part of an animal's life cycle or for a particular season, whereas another habitat is needed during another time of year. The most common migrations observed among frog species are annual movements, typically in the spring, from hibernation sites to aquatic breeding sites, where males and females congregate. After mating, both sexes in most cases move to feeding areas for the spring and summer and return to their hibernation sites for winter. Some frogs, such as spadefoot toads that live in underground habitats where they remain dormant for months at a time, may move to breeding ponds after heavy rains at any time, but this is still a form of migration as the individuals eventually return to their underground burrows. Some treefrogs spend much of their life in treetops and will literally jump to the ground to migrate down to breeding sites and then return later by climbing back up the tree.

Many species follow the same migration routes year after year, a trait that can create real problems for a species when a highway is built between their breeding sites (e.g., wetlands) and other habitats where the species spends its time. Roads can be particularly problematic for species such as wood frogs or spadefoot toads that have mass migrations in which thousands of individuals move over a short period. Highway mortality of frogs can be especially high under certain conditions of timing and location, resulting in significant tolls on some frog populations.

Some individual frogs have been recorded migrating long distances for purposes other than breeding. For example, Jim Dole documented that most northern leopard frogs (*Rana pipiens*) in a Michigan population moved less than 30 feet per day during the summer, except during nighttime rains, when some individuals moved as much as 250 yards, later to return to their original locations. Some studies suggest that leopard frog migrations from wetlands to hibernating sites during autumn may be more than 2 miles.

Most frogs migrate at least short distances for breeding purposes, but they do not necessarily migrate back to the same wetland in which they hatched and metamorphosed. Instead, they may disperse from their natal site to another habitat where they spend most of their time and then move to a different breeding site.

Susumu Ishii and colleagues from the University of Tokyo examined a population of Japanese toads (*Bufo japonicus*) and speculated that toadlets leaving the wetland for the first time use their sense of smell to become

The striped burrowing frog (*Cyclorana alboguttata*) from Australia spends most of its time underground when conditions are dry but emerges and migrates to temporary pools to breed during wet weather.

Courtesy H. B. Hines DERM

familiar with the route. Then, as much as 3 years later when they return to breed, they use local olfactory cues to follow the same route.

How do frogs survive the winter?

For most frogs living in tropical or subtropical regions, winter is not a problem (i.e., there is no winter). However, for those living in the Temperate Zones, some even extending as far north as the Arctic Circle, protection from winter cold is essential for survival. Frogs accomplish overwintering or winter dormancy, also referred to as *hibernation*, or *brumation*, in various ways. Adults of many species avoid freezing temperatures during winter by selecting hibernation sites that do not freeze. Some overwinter on land, buried beneath vegetation or in burrows, and some stay under water and bury themselves in the mud at the bottom of ponds. Because a frog's metabolism is low at cold temperatures, the oxygen requirements are low, whereas the oxygen levels in cold water are relatively high. Consequently, many frogs are able to survive under water during the winter with only the available dissolved oxygen that diffuses across their skin from the water. Some species, including bullfrogs and green frogs, overwinter as tadpoles in the muddy bottom of lakes or ponds.

Another mechanism some frog species use to deal with cold winter temperatures is to produce a form of antifreeze in the body that allows them to survive temperatures several degrees below freezing. These species still seek out areas that protect them from severe cold, but they live in cold climate regions, where winters can be especially harsh so that antifreeze measures are highly adaptive. Most frogs that have been studied produce glucose or urea in their blood and other tissues as protection from freez-

Frogs: The Animal Answer Guide

Some Temperate Zone frogs such as the wood frog (*Rana sylvatica*) have antifreeze products in their blood and other tissues to provide protection from temperatures below freezing. The geographic range of wood frogs extends north of the Arctic Circle. Courtesy John D. Willson

Cascades frogs (*Rana cascadae*) move to breeding sites during cold weather before snow has melted, as seen in this individual from Mt. Ranier, Washington, USA.
Courtesy Cris Hagen

ing temperatures below 30 degrees Fahrenheit and some as low as 20 degrees Fahrenheit. Many of these species actually partially freeze. Water in-between their cells (i.e., extracellular water) can freeze, and some species, such as wood frogs, can withstand freezing of up to 70 percent of their total body water. The common gray treefrog (*Hyla versicolor*) has been reported to avoid freezing by using glycerol in its tissues as an antifreeze instead of glucose.

How do frogs survive droughts?

Species of frogs that live in regions subject to drought have evolved mechanisms to persist for long periods without rain. The primary approaches used by frogs to avoid dehydration are to avoid contact with the dry air as much as possible by hiding under leaves, ground debris, and other moist areas or retreating into underground burrows. They remain inactive much of the time, and some species, such as spadefoot toads, are known to be able to survive for months underground. African bullfrogs (*Pyxicephalus adspersus*) go underground during extended dry periods of the year and form a cocoon to minimize moisture loss. If they emerge, it is generally only at night when temperatures are cooler and humidity higher. Some toads in the family Bufonidae rely on long-term water retention and have relatively impermeable skin that minimizes water loss, even over long periods. Some species solve the drought problem by moving overland to aquatic sites that may not be as productive as their home site but are more permanent. However, such long-distance moves are usually made during infrequent rains during an extended drought. Some species of leaf frogs (genus *Phyllomedusa*) of South America secrete a fatty substance that they spread over their body to make a waxy film that reduces water loss.

Frog populations in a region fluctuate dramatically in numbers over time, and droughts often substantially reduce population numbers. However, the ability of frog populations to recover can be equally dramatic. For example, during a 3-year drought in South Carolina, virtually no frogs laid eggs that produced tadpoles and metamorphs at a wetland. In the fourth year, rains were frequent, and the weights of the frogs emerging from the wetland were taken. In one season, the wetland produced more than 3 tons and 300,000 juveniles of 17 species of frogs and more than 1.6 tons of amphibian biomass.

Some species rely on cannibalism to respond to drought conditions in which favorable breeding conditions are infrequent and short-lived. Tadpoles of spadefoot toads often cannibalize others of their own species in a rapidly drying rainwater pools so that they can metamorphose more quickly and get out of the pool before it dries.

Do frogs have enemies?

Frogs have many enemies, most of which intentionally kill them for food. Other enemies, namely, humans, may cause harm unintentionally, but can be much more damaging to frog populations or even entire species through the destruction and alteration of their habitats. Adult frogs have many natural predators, including fish, birds, snakes, turtles, mammals, and other frogs. Fish, snakes, and turtles commonly prey on tadpoles, as do some of the larger aquatic invertebrates, such as dragonfly naiads (larvae). A variety of parasites and diseases, such as chytrid fungus, are natural enemies of many species of frogs. Frog eggs are particularly vulnerable to predation from numerous aquatic insects, fish, and salamander larvae. Most frog eggs never make it to the adult stage.

Unfortunately, people have become a primary enemy of frog species throughout the world. Each time a highway is built near a wetland habitat, thousands or even millions of frogs are destined to die while trying to cross the road. The resulting habitat fragmentation has additional negative impacts. Introducing predatory fish such as largemouth bass or various species of trout into waters where they do not occur naturally ensures the elimination of most native frog species. Pollution, such as toxic chemicals, acid rain, or heavy metals, can cause malformations or mortality and eventually causes populations to decline and become extinct. Finally, one of the greatest negative impacts on frog populations everywhere is the destruction of small wetlands. Collectively, these human-caused impacts have reduced the numbers of frog populations in most parts of the world.

How do frogs avoid predators?

Avoiding predation is critical for frogs in every life stage because they have no protective armor and are easy prey for many animals. Remaining inactive in a hiding place, a common behavior pattern to an ectothermic lifestyle, is one of the most common forms of protection for frogs. Many

A common predator of toads (genus *Bufo*) in the southeastern United States is the eastern hognose snake (*Heterodon platirhinos*). Courtesy John D. Willson

A small green treefrog (*Hyla cinerea*) falls prey to a large, carnivorous insect, a praying mantis. Courtesy John Mackay

species are inactive for most of their lives, making them nearly inaccessible to most predators. Frogs are usually active at night when relatively fewer predators, such as birds, are hunting. Even when some frogs are calling, they are exceedingly well camouflaged or hidden in vegetation. Anyone who has attempted to find a little grass frog (*Pseudacris ocularis*) calling from less than 3 feet away can attest to the effectiveness of their abilities to go unseen. A few species, such as the poison frogs (family Dendrobatidae), are visible and active during the day because their aposematic coloration warns predators that they are poisonous. Of course, most frogs use jumping to escape from predators. See chapter 4 ("Frog Behavior") for details on how frogs defend themselves.

Do frogs get sick?

Frogs are susceptible to many diseases and parasites, including chytrid fungus, red-leg, ranaviruses, and trematode parasites. Chytrid fungus is described in detail in chapter 10. Red-leg, or *Aeromonas*, is commonly observed in captive frogs or toads that have been handled too much. The most obvious expression of red-leg is the presence of blood vessels on the inside of the thighs or on the belly of the animal that cause these normally white regions to appear red. Redness usually indicates internal hemorrhaging caused by the *Aeromonas* bacteria. Ranaviruses, or iridoviruses, can be lethal to frogs or tadpoles of some species, and the virus can be readily transmitted to other individuals in a pond. Trematode parasites, which use

Southern cricket frog, *Acris gryllus*, southeastern United States. Courtesy Aubrey M. Heupel

Broad-headed rain frog, *Craugastor megacephalus* La Selva, Costa Rica. Courtesy John D. Willson

Carolina gopher frog, *Rana capito*, southeastern United States. Courtesy Thomas Luhring

Chilean mountains false toad, *Telmatobufo venustus*, Chile and Argentina. Courtesy Dante Fenolio

Clown treefrog, *Dendropsophus leucophyllatus*, Peru. Courtesy Chris Gillette

Convict treefrog, *Hypsiboas calcaratus*, Peru. Courtesy Chris Gillette

Coronated treefrog, *Anotheca spinosa*, Central America. Courtesy Dante Fenolio

Darwin's frog, *Rhinoderma darwinii*, Villarrica National Park Chile. Courtesy Dante Fenolio

Eastern spadefoot toad, *Scaphiopus holbrookii*, South Florida. Courtesy Chris Gillette

arlequin frog, *Atelopus pulcher*, Peru. Courtesy Chris Gillette

Holy Cross toad, *Notaden bennettii*, Australia. Courtesy H. B. Hines DERM

Litter toad, *Bufo haematiticus*, La Selva, Costa Rica. Courtesy John D. Willson

Little grass frog, *Pseudacris ocularis*, southeastern United States. Courtesy Thomas Luhring

This large, arboreal frog (*Platypelis grandis*) is found in eastern Madagascar and has no regularly used common name. It is a member of the family Microhylidae and often hides in tree holes to avoid predators. Courtesy Leslie Ruyle

frogs as a host species, are suspected of causing malformations in frogs from numerous locations in the United States. A common parasite identified as causing missing limbs and other deformities in frogs also causes "swimmer's itch" in humans.

Stress is believed to make frogs susceptible to disease, and humans probably lower frogs' resistance to disease and create stress through various forms of pollution. Joseph M. Kiesecker of Pennsylvania State University provided convincing evidence that trematode parasites can cause limb deformities in wood frogs (*Rana sylvatica*) and that the problems were exacerbated when pesticide runoff was present. Changing climate or the quality of ultraviolet radiation from sunlight may also trigger complications in frogs exposed to diseases that might be unproblematic in healthy, unstressed frogs. Even wetland use by cattle may cause stress for tadpoles. The interaction between chytrid fungus and pollution or climate change has been proposed as a potential problem for many species of frogs in Central America.

How can you tell if a frog is sick?

Often it is not easy to tell whether a frog is sick. If a frog is lying on its back, is completely listless, or is thin and emaciated, it is probably sick. Red-leg usually indicates that a frog is not doing well. Determining exactly what makes a frog sick can be difficult, and frogs often die before an accurate diagnose can be generated.

Are frogs good for the environment?

Frogs are extremely important components of many natural ecosystems and play key roles as both predators and as prey for other animals. They can occur in extremely high densities in some habitats and, thus, can account for an incredible amount of biomass. All frogs are carnivores as adults (i.e., they eat other animals), and their role as predators is vital because most frogs eat insects, including human pests. The extent to which they actually control insect pests, such as mosquitoes, is difficult to determine, but their potential for controlling insect populations is likely substantial. Frogs and toads are also essential as prey for many species of animals. Many species of birds, snakes, mammals, and other animals feed on frogs either in their adult stage or as tadpoles. Some species of snakes are so specialized they eat only a few species of toads their entire lives.

Chapter 6

Reproduction and Development

How do frogs reproduce?

All frogs have two sexes, females that produce eggs and males that produce sperm. Thus, all species reproduce sexually, with the male's sperm fertilizing the female's eggs. In most species, the female deposits her eggs outside of her body before fertilization. In the simplest situation, eggs are released from the female's body into the water, and the male releases sperm on the eggs at which time fertilization occurs. During mating, the male in nearly all frogs grasps the female around the waist or under the arms, a process called *amplexus*, so that their cloacas are in close juxtaposition when the eggs and sperm are released. Because an egg and a sperm each has half the number of chromosomes as each animal, the resulting embryo has the full chromosome complement from the egg and sperm.

Most frogs have somewhat elaborate mating procedures that begin with male frogs calling to attract females and then males and females congregating at a breeding site. Although males establish territories to compete for mates and sometimes engage one another in physical combat, the female generally chooses her mate before she releases her unfertilized eggs. Eggs usually hatch within a short time and, in many species, have free-swimming tadpoles that eventually emerge as the adult form. Although the common perception in North America and Europe is that the tadpole stage is typical of frogs, a high proportion of the world's frog species undergo complete larval development in the eggs and hatch as froglets. Direct development of this sort is the norm for many tropical species of frogs. For example, none of the more than 180 species in the genus *Eleutherodactylus* have tadpoles.

A male wood frog (*Rana sylvatica*) clasps the larger female as he waits to fertilize the eggs when she releases them into the water. Courtesy Victor Lamoureux

Do all frogs lay eggs?

Females of all frog species produce eggs, and most, but not all, lay them in an unfertilized state in the presence of a male that fertilizes eggs outside the female's body. Most people are familiar with frogs that lay their eggs in water, where they hatch and develop into tadpoles. However, many species of frogs, especially in the Tropics worldwide, lay their eggs in moist places on land, and juvenile frogs emerge directly from the eggs without a free-swimming tadpole stage.

At least five species of frogs are live-bearers, meaning that they produce eggs fertilized inside of the female's body where they develop without being laid. Several species are in the genus *Nectophrynoides* (family Bufonidae) of Tanzania and a single species, the golden coqui (*Eleutherodactylus jasperi*) of Puerto Rico, is in the family Leptodactylidae. Depending on the species, the gestation can last from 1 to up to 9 months. The developing embryos are nourished inside the body of the female, who eventually gives birth to fully developed little froglets. Marvalee Wake of the University of California at Berkeley studied the golden coqui and determined that egg yolk was the only evidence of provisions by the female to the embryos. Egg yolk was still present when the baby frogs were born, and no direct maternal nourishment was given by the female. Additional species of frogs with internal development of eggs and embryos probably exist but have not yet been discovered and described.

How many eggs do frogs lay?

As with most other groups of animals with numerous species, the number of eggs produced by a female varies and depends on the species. Some frogs and toads are known to lay enormous numbers of eggs that may be in strings or clusters that number in the hundreds or thousands. In contrast, some frogs lay fewer than 10 eggs at a time. Water-breeding species lay the

Frogs: The Animal Answer Guide

The Kihansi spray toad (*Nec-tophrynoides asperginis*) of Tanzania is on the verge of extinction in the wild and is one of the few species of frogs in the world to have internal fertilization and bear live young.

Courtesy Dante Fenolio

largest numbers of eggs. American bullfrogs (*Rana catesbeiana*) can lay up to 20,000 eggs at a time. The cane toad (*Bufo marinus*) has been known to lay more than 30,000 eggs at a time.

Of course, offspring from most eggs laid by any species of frog do not survive to adulthood, and, in many situations, none survive during a particular year or at a particular location. For example, a drought may cause wetlands to dry, thus killing all eggs of a particular species in a particular year. Consequently, frog population sizes characteristically decrease in some years and increase in others, but not necessarily because of variation in the number of eggs laid. Fluctuations in population sizes of frogs can be dramatic. In situations in which a frog species is not declining, an average female frog will replace herself and her mate during her lifetime with two offspring (a male and a female) that survive to sexual maturity. So, for a natural frog population that remains stable over time, the lifetime average number of surviving offspring will be the same for species that lay thousands of eggs as it is for those that lay only 10.

Where do frogs lay their eggs?

Where frogs lay their eggs depends on the habitat in which they live and the frog species. Most people are familiar with frogs that lay their eggs in

Some aquatic breeding frogs, such as these wood frogs (*Rana sylvatica*), may lay hundreds or even thousands of eggs in large, jellylike clusters in shallow water. Courtesy Victor Lamoureux

puddles, ponds, or lakes where they turn into aquatic tadpoles before meta-morphosing into frogs or toads. This is the egg-laying pattern for most frogs in the Temperate Zones of North America and Europe. The Pacific tailed frog (*Ascaphus truei*) and Rocky Mountain tailed frog (*Ascaphus montanus*) differ from other North American frogs that lay their eggs in water because they deposit eggs in fast-flowing streams. Special adaptations are necessary for the eggs and tadpoles not to be washed away. When tailed frogs mate, sperm is deposited in the female rather than the female depositing eggs directly into the water for the male to fertilize. Mating usually occurs in the fall, and the female retains the eggs until spring, so that when she releases the eggs into the water, they are already fertilized. The female lays strings of eggs beneath large rocks that protect them from strong currents. Tadpoles are equipped for a stream environment. Large suction-like mouths allow them to cling to rocks as they feed on algae.

Despite the perception among many people that frogs lay eggs in water, the diversity of other microhabitats where they lay their eggs and where eggs develop is astounding, especially in the Tropics. Some tree-climbing frogs attach their eggs to the underside of tree leaves overhanging water, positioning the nesting site so that hatching tadpoles fall into water. Some frogs lay their eggs, or often only a single egg, in little puddles that form in the axils of bromeliads or other plants in the rain forest. Several burrowing species lay their eggs underground, and some ground-dwelling forest species lay their eggs in wet, dead leaves or other vegetation. Many species in the genus *Eleutherodactylus* that have direct development, in which a froglet emerges from the egg with no aquatic tadpole stage, lay their eggs in various terrestrial sites, including beneath rocks and logs, in the dry axils of plants, or sometimes on the ground. The female coqui (*Eleutherodactylus coqui*) deposits eggs that have been fertilized internally in areas with hidden

Frogs: The Animal Answer Guide

One of the species of marsupial frogs, *Gastrotheca cornuta*, from tropical America in which the female carries the eggs in a pouch on her back. Courtesy Dante Fenolio

The salmon-striped frog (*Limnodynastes salmini*) is found in eastern Australia and constructs large foam nests where it lays its eggs. Courtesy H. B. Hines DERM

spaces, including leaves or palm fronds that have fallen to the ground and rolled up as they dried. The male guards the eggs until they hatch by direct development.

The Surinam toad (*Pipa pipa*) lays its eggs in an unusual way. The male and female perform underwater acrobatics in which the female releases the eggs and the male releases sperm to fertilize them. During the mating ritual, the male places the eggs on the female's back. Later, her skin grows over the eggs, where the larvae develop and emerge as froglets.

How do frogs protect their eggs?

Frogs protect their eggs from numerous threats, especially predation and dehydration, in a variety of ways. These can include carefully selecting the site where the eggs are laid and where the tadpoles will develop, nest guarding, and laying eggs on vegetation above the ground. The mortality rate of frog eggs is high, and some species, such as American bullfrogs and marine toads, compensate by laying exceedingly high numbers of eggs so that even when only a small percentage survive, stable population levels are maintained. In some species, such as the gladiator frog (*Hypsiboas rosenbergi*) of Central America, the male stays with the eggs, which are deposited in mud nests, and will defend them against predators and other males. Adults of the treefrog *Chiromantis hansenae* of Thailand sit on the eggs, more than 200 of which are deposited on leaves or rocks so that the hatching tadpoles fall into water below. Hatching occurs in less than a week, and the attending adult is present more than 70 percent of the time. Their presence presumably prevents the eggs from drying and may protect them from parasitic flies and fungal infections.

Some species carefully select egg-laying sites that will afford the greatest overall protection, including underground, in the axils of leaves in trees, and on dead vegetation on the ground. An excellent example of underground protection is that of the African frogs in the genus *Hemisus* (including the spotted snout-burrower, *Hemisus guttatus*; shovel-nosed frog, *Hemisus marmoratus*; and olive snout-burrower, *Hemisus olivaceus*). These species burrow during their normal life cycle, spending much of their life underground. During the mating process, the female digs a burrow near a wetland where the soil is moist, and the male fertilizes the eggs as she lays them underground. Burrows may be more than 8 inches below the ground's surface. To avoid desiccation, the female deposits a layer of jelly-like material atop the eggs. She then sits in the burrow above the jelly layer and eggs for several days until they hatch. When tiny tadpoles emerge, the female digs a tunnel toward the wetland and eventually connects with water, and the tadpoles begin a free-swimming period. The elegant frog (*Cophixalus concinnus*) of Queensland, Australia, also does not lay its eggs in water but hides them beneath rocks or logs where soil is moist. Eggs develop directly into froglets without a tadpole stage.

Some glass frogs (genus *Hyalinobatrachium*) deposit their eggs on the undersides of leaves as the male fertilizes them. The male returns at night to empty its bladder on the eggs, which protects eggs from dehydration. This process known as *hydric brooding* has also been observed in the strawberry poison frog (*Dendrobates pumilio*) in Costa Rica. In this species, the male keeps the eggs moist for a week or more by periodically emptying his

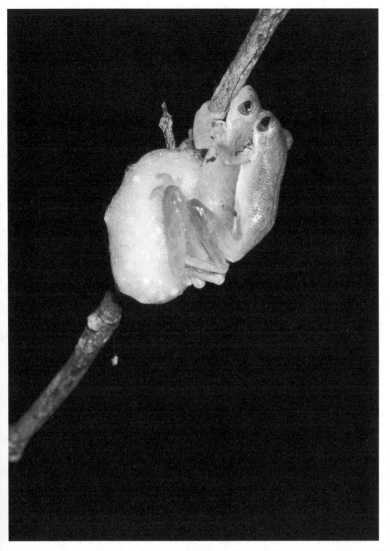

Some species, such as the moss frogs (family Rhacophoridae) from Southeast Asia, lay their eggs on vegetation above water and build foam nests to avoid dehydration of the eggs. Courtesy Wayne Van Devender

bladder over them. When the eggs hatch into tadpoles, the female transports them on her back to the axil of a bromeliad leaf or other plant, usually one egg per leaf axil that has collected water. Only one tadpole survives per axil, and the female deposits unfertilized eggs for the tadpole to eat.

Another strategy for avoiding dehydration is foam nests, which have been reported for dozens of genera and several different families of frogs. In some species, such as the arboreal gray-eyed frog (*Opisthothylax immaculatus*), the only foam nest builder in the family Hyperoliidae of equatorial Africa, both the male and female are involved in making the foam nest. Perhaps the ultimate in egg protection can be seen in the large (greater than 8 inches) frogs known as mountain chickens (*Leptodactylus fallax*) found on Dominica and Montserrat in the Caribbean. On the basis of studies by Richard C. Gibson and Kevin R. Buley, both sexes engage in a variety of behaviors, including foam nest building, that protect eggs from dehydration, predation, and starvation. During mating, the female deposits eggs in a terrestrial burrow, and after fertilizing the eggs, the male kicks the viscous fluid surrounding them into a froth over a several-hour period. A day later, the layer surrounding the foam nest dries and becomes resilient enough for the female to sit atop the eggs. She stays in the burrow, and the male leaves, usually staying within 3 feet of the burrow entrance. The foam prevents the eggs from drying out, and either sex will defend the eggs from potential predators. When eggs hatch, tadpoles remain in the burrow, even though there is no water. During the approximately 45 days, the female feeds them directly by depositing unfertilized eggs for them to eat and continues to provide foam for the nest. The number of tadpoles generally ranges from 26 to 43, but the female may lay as many as 25,000 unfertilized eggs to feed them.

Frogs have numerous other strategies for protecting their eggs, including brooding eggs in their mouth or stomach or in back pouches. The Surinam toad (*Pipa pipa*) mentioned earlier is a completely aquatic species that breeds in the water. The male places the eggs on the female's back. Her skin grows over the eggs, which remain in place until they hatch. Three species of frogs actually protect their eggs by consuming them. In two species of South American frogs in the genus *Rhinoderma*, the male picks up the developing eggs in his mouth; they continue to develop in his vocal sacs. In the Chile Darwin's frog (*Rhinoderma rufum*), embryos develop into tadpoles, and thereafter, the male hops to the water and deposits them as free-living tadpoles. In the other species, Darwin's frog (*Rhinoderma darwinii*), after the male picks up the eggs (up to 19 have been documented), he lets them develop into tadpoles that remain in his vocal sacs for up to 70 days. Tadpoles eventually metamorphose and emerge from the male frog's mouth as froglets.

Another remarkable case of parental care among frogs is that of the two species of gastric-brooding frogs (genus *Rheobatrachus*) in Australia, both of which are believed to be extinct. Females of both species pick up the developing eggs in their mouths and incubate them in their stomach. Physiological changes occur in the female's stomach so that all digestive processes are turned off during the egg-brooding period. Females of the southern gastric-brooding frog (*Rheobatrachus silus*) can hold as many as 25 developing young in the stomach for up to 7 weeks. The last reported specimen of this species was found in 1981. Only one birth of 22 froglets was documented for the northern gastric-brooding frog (*Rheobatrachus vitellinus*). The last individual of this species was seen in 1985.

Do frogs care for their young?

Frogs care for their young in many ways, including some impressively elaborate forms of parental care. The male Darwin's frog broods the young in their vocal sacs, and the female gastric-brooding frogs of Australia broods her young in her stomach. Some poison frogs in the genus *Dendrobates* carry their tadpoles on their back to deposit them in the water. According to Marty Crump of Northern Arizona University, this form of parental care, involving tadpole transport, has been documented in no fewer than seven separate families of frogs. The South American frog, *Osteocephalus oophagus*, exemplifies another type of parental care practiced by many tropical species. The female of this species lays her eggs in water-filled axils of leaves and returns every few days to deposit unfertilized eggs, which the tadpoles eat.

A common phenomenon among animals, including frogs, is that species that lay extremely high numbers of eggs (in the hundreds or thousands) are far less likely to provide any parental care to their eggs and young than those that have smaller numbers. That is, frogs may invest in their offspring by producing large numbers of eggs of which a tiny few are likely to survive, or they can produce fewer eggs but invest energy into maximizing the likelihood that those few eggs will survive and become young frogs. No frogs are known to care for their young beyond the tadpole stage.

Why do some frogs lay their eggs in strange places?

Many species of frogs lay their eggs in "strange" places, but their goal is to maximize the likelihood of successful reproduction from the egg through the tadpole stage. Many unusual nesting patterns and strategies of frogs have been mentioned in previous questions. Many frogs, especially those that lay only a few eggs, select hidden places not visible to potential predators. Many animals, both vertebrates and invertebrates, will readily eat frog eggs, so making them difficult to find is strategic. Some snakes in the Tropics nearly exclusively eat frog eggs. Some frogs select places, such as underground burrows or the underside of leaves, so that the adult can remain nearby to protect eggs from small predators.

Do all frogs have a tadpole stage?

In Temperate Zone regions of Europe and North America where biological studies on frogs and toads have been conducted, nearly all frog species go through the typical free-swimming tadpole stage with which most

Fleischmann's glass frog (*Hyalino-batrachium fleischmanni*) of tropical America lays its eggs on the underside of leaves where they are hidden from predators. Courtesy Jonathan Campbell

people are familiar. However, many tropical species do not have the aquatic tadpole stage, often considered a common trait of frogs and toads.

All frog species have eggs that develop into larvae, but in many, eggs are laid on land, and the larvae later emerge as tiny frogs after undergoing complete development inside the egg without a free-swimming tadpole stage. A few species even develop completely in the egg while still in the reproductive tract so that young frogs are live-born directly from the mother. In more unusual frog species, eggs are released from the female and fertilized but then are taken into the mouth or stomach or placed in pouches on the back and later emerge as froglets.

How many baby frogs grow to be adults?

The simple answer is that a small proportion of the eggs laid by frogs and toads every year develop into tadpoles, and a small proportion of tadpoles survive to metamorphose as froglets or toadlets. Of the survivors that do end up on land, only a small number ever reach adulthood. In a population maintained by a particular frog species in a region over time, survivors that reach adulthood and successfully reproduce are roughly equal to the number of adults that die. Thus, for a species such as American bullfrogs or marine toads that lay tens of thousands of eggs, only a tiny fraction grow to adulthood, and the eggs laid by many female frogs have 100 percent mortality.

How fast do frogs grow?

Determining the rate of growth in body size of any animal species is a critical part of ecological studies. In addition, frogs have complex life cycles in which a larval or tadpole stage is involved between the time an

Frogs: The Animal Answer Guide

Frogs and toads native to North America, such as the pine woods treefrog (*Hyla femoralis*), have an aquatic tadpole stage but many tropical species have direct development of their eggs in terrestrial situations.

Courtesy Aubrey Heupel

egg is fertilized and a juvenile frog is produced. Determining the rate at which each stage is reached is another measure of growth of the individual. Many variables should be considered when determining the growth and development rates of tadpoles, especially the species, food availability, temperature, and number of predators present. Tadpole density—the number of tadpoles per volume of water in the habitat—can influence the rate at which some species metamorphose. This factor is important for many species of true toads (genus *Bufo*) and spadefoot toads (genus *Scaphiopus*) that breed in temporary, rain-filled puddles that may dry up within a few days under certain conditions. Toad tadpoles in a rapidly drying aquatic area will often metamorphose days earlier, and at a smaller size, than those in a puddle where water levels remain constant. This plasticity in development rate characterizes many frog species whose tadpoles confront unpredictable environmental vagaries.

In North America, spadefoot toads have the shortest larval period of any frog species—the tadpole stage from hatching to metamorphosis. Spadefoot tadpoles in a fast-drying pool or puddle have been documented to emerge from the water's edge as toadlets between 1 and 2 weeks after hatching. An excellent example of the plasticity in larval development of the species, and perhaps of frogs in general, can be seen in the studies conducted on tadpoles of Couch's spadefoot toad (*Scaphiopus couchii*) by Robert A. Newman in Big Bend National Park, Texas. Newman conclusively demonstrated that the larval period can depend on the rate at which breeding pools dry after eggs are deposited. In pools that retained water and adequate food for the tadpoles, metamorphosis continued for 16 days; in pools that dried most rapidly, tadpoles metamorphosed into toadlets within

8 days. However, the body size of emerging toadlets was directly correlated with time of metamorphosis. Thus, spadefoots emerging after 8 days were generally smaller than those emerging after 12 days, and these would be smaller than those emerging after 16 days. Because food availability and temperature influence such phenomena, categorical statements must be made cautiously. In general, many frogs probably have a range of flexibility of how rapidly they metamorphose, depending on environmental circumstances. Hence, precise numbers about the rate of metamorphosis are not practical. In contrast to the rapid larval development of spadefoot toads, metamorphosis in some species, such as the green frog (*Rana clamitans*) and bullfrog, has been known to take as long as 2 years, which means tadpoles spend at least one or two winters in the water. Because they can grow quite large as tadpoles, they are larger when they metamorphose and, thus, have a higher likelihood of surviving and reproducing as adults.

Growth rates of individuals from the froglet/toadlet stage to adulthood vary considerably among different species and are influenced by numerous variables. For species with aquatic, free-living tadpoles, the larval stage until metamorphosis influences the size at which juvenile frogs emerge from the water. Smaller juveniles in some instances do not grow as rapidly or have higher mortality rates than those that begin life at a significantly larger size. Juveniles in all species grow rapidly until maturity is reached at which time growth slows considerably and presumably eventually stops. Most species of frogs that have been studied in the field grow to adult size and reach maturity within 1, or at most 2 years, after metamorphosis. Growth rates of frogs are typically measured in millimeters per year and are higher in species that have large adults than in those with smaller ones.

How is the sex of a frog determined during development?

Frogs have sex chromosomes that determine their sex, although in contrast to humans in which females always have two X chromosomes and males have a single X chromosome and a shorter Y chromosome, the sex chromosomes in frogs are the same size in both males and females but differ genetically. The specifics of genetic sex determination are complex in frogs and can even differ within the same species living in different localities, but in general, a pair of chromosomes determines an individual's sex.

Some species of frogs make matters even more complicated because an individual can change sex—a male can become a female or vice versa. For example, both sexes in toads in the family Bufonidae have a small structure known as the Bidder's organ connected to the ovaries of females and the testes of males. Although the organ's basic function remains unknown, experiments with toads have demonstrated that, if the testes are removed or irreparably damaged, the Bidder's organ develops into a functioning ovary and the individual becomes a female. A complex system of sex reversal has also been reported in the common frog (*Rana temporaria*) of Europe in which individuals have ovaries that develop into testes in certain situations, such as when a female's ovaries become damaged. The mechanisms underlying sex reversal remain a mystery, and it is likely that, following more careful study, other species of frogs will be found to have complex factors that determine an individual's sex as an adult.

How can someone tell a frog's sex?

The sex of adults of some species of frogs can be determined by dissecting the animal to determine whether ovaries or testes are present. However, many external clues also exist that can be used to classify a frog as either a male or a female. The sex of tadpoles or juvenile frogs can only be identified by dissection or genetic analysis and, in many cases, cannot be easily determined.

A reliable method for identifying the male in which the male has a single vocal sac is to examine the throat during the breeding season. It will generally have a darker coloration than the female. With practice, an adult of either sex can be easily determined. In many species, such as the true toads (family Bufonidae) and the spadefoot toads (family Scaphiopodidae), the sex of adult females can often be readily recognized because of their much larger size. The male is larger in some species, such as the American bullfrog, so that size can be used to differentiate between the sexes in some situations. Males of the leopard frogs and many other species in the genus *Rana* have an unusual, readily recognizable trait in which the thumbs on their forelimbs become enlarged during the mating season. This enlarged thumb is used to hold onto the female during mating. The size of the tympanum is noticeably larger in males of some frog species and can determine the frog's sex if both sexes are compared. In a few frog species, males and females are different colors. The male golden toad (*Bufo periglenes*), which is apparently now extinct, is a solid, strikingly bright orange color, whereas the larger female is greenish or even black with only a few reddish spots.

In most frog species worldwide, highly recognizable advertisement calls characterize males but not females. Also, a frog's behavior can sometimes reveal its sex. For example, males often arrive at breeding ponds before females, so on the first days of calling, nearly all individuals are males. Also, when frogs or toads that mate in water are in amplexus, the male is typically on top.

The thumbs on the front feet of male leopard frogs in the genus *Rana* become enlarged during the mating season. Courtesy Thomas Luhring

How can you tell the age of a frog?

Frogs have no obvious features, such as the rings of a tree or growth rings on turtle shells, to indicate the age of a living individual. However, the state of development or body size can sometimes indicate age. A tadpole of a particular species is likely to be younger than individuals of the species that have already metamorphosed and are living on land. Likewise, frogs grow continually after metamorphosis to their adult size over a period that usually ranges from a few months to 1 or possibly 2 years. A smaller frog of a particular species and sex (one sex is often larger than the other in some species) is usually younger than a larger individual. However, such age estimates must be done with caution as the size of recently metamorphosed froglets or toadlets can vary greatly, depending on environmental conditions they experienced as tadpoles. Once frogs reach adult size, they may live for several years, and determining one's age based simply on its appearance is not possible.

Some amphibian biologists have used a technique known as *skeletochronology* in which a cross section of a leg bone of a preserved frog or the toe of a live one is prepared so that circular rings called growth rings can be counted. Narrow rings that represent periods of slow growth are concentric; wider rings are presumed to have occurred during periods of rapid growth. Thus, in species with periods of activity and growth during spring and summer, which are distinct from periods of dormancy (fall and winter) and minimal growth, the number of rings should indicate the number of years the animal has lived. The concept is analogous to counting the rings of trees to determine age. However, some scientists question the validity of the technique in accurately estimating the age of older amphibians because

Frogs: The Animal Answer Guide

growth slows considerably after maturity, and the rings become closer together and more difficult to differentiate. The technique is probably more accurate with some species and in some regions than with others because of the length of the growing season and the possibility of false growth rings occurring as a result of intermittent feeding throughout the year.

How long do frogs live?

Because determining the precise age of a frog or toad is difficult or impossible without knowing when the individual was born, species records for known-age frogs are based either on captive-raised animals acquired as juveniles or on field studies in which marked animals were recaptured over time. Both approaches can lead to biased interpretations in which the maximum age of individual frogs is either longer or shorter than can be measured.

In field studies, most frogs would be continually subjected to predation or extreme environmental conditions that could shorten the potential life span of individuals. However, the odds of recapturing a particular marked frog over extended periods of several years is low, even if the frog is still alive, leading to an underestimation of individuals' life span. Records of frogs' life spans based on field mark-recapture studies depend on the longevity of the study. An individual frog cannot be determined to be much older than the time between its first and last capture. Consequently, field studies that have measured maximum longevity of species are limited.

Captive-raised animals can be especially informative in determining the maximum potential longevity of a species because mortality resulting from most of life's natural hazards, such as predation, drought, or extreme cold, can be avoided. Thus, a pet frog that has lived for many years documents maximum physiological longevity, although such ages might rarely or never be achieved in the wild. Some animals, including many species of frogs, can be difficult to raise in captivity so that their true life span potential in the wild is not realized. Longevity records of captive animals often underestimate the potential life span of the species because the individual was a wild-caught specimen that might already have been several years old. Longevity records of some species are based on individuals that are still alive, which means the maximum longevity is still to be achieved.

Excellent records exist for the ages of captive frogs because of the publications by Frank Slavens and Kate Slavens who inventoried the longevity of reptiles and amphibians in captivity. The inventory, which was based on submitted records by pet owners and zoos throughout the world and from published literature, first appeared in print in 1976. From 1980 through 1999, the inventory was published annually, and the records through 1999

Green poison frogs (*Dendrobates auratus*) have been documented to live more than 20 years in captivity.
Courtesy John D. Willson

can be found at www.pondturtle.com/lfrog.html. The final installment of the frog longevity inventory included records for 16 families, 59 genera, and more than 180 species of frogs and toads.

When the Slavens retired, many of the individual frogs on which records were based were still alive, but the inventory was not continued after 2000. Available records indicate that some frogs potentially have long life spans. The longest was a European toad (*Bufo bufo*) that lived for 40 years and another for 36. That 24 genera and 51 species, representing 12 families, were documented to have had captivity records of more than 10 years demonstrates that frogs can potentially have extended longevity and suggests that many species may live for long periods in the wild. An additional indicator that frogs probably live for many years is that examples of those documented as living a decade or more represented all six warm continents as well as Madagascar and Puerto Rico.

Some notable longevity records for frogs in captivity, in addition to the European toad, are the African clawed frog (*Xenopus laevis*), 30 years; Blomberg's toad (*Bufo blombergi*) from Colombia, South America, 28 years; cane toad (*Bufo marinus*) from tropical America, 24 years; and European treefrog (*Hyla arborea*), 22 years. Other species reported to have lived at least 20 years in captivity are the European fire-bellied toad (*Bombina bombina*), green poison frog (*Dendrobates auratus*), White's treefrog (*Litoria caerulea*), and Cuban toad (*Bufo peltocephalus*).

Because of the difficulties in obtaining age and longevity records for frogs in the wild, relatively few records are available. G. J. Measey and Richard C. Tinsley conducted a mark-recapture study that revealed that African clawed frogs in a feral population in South Wales, United Kingdom, could live at least 14 years in the wild, even outside their native habitat. Although these frogs are not native to the area, the study augments the

Frogs: The Animal Answer Guide

captivity data, indicating individual African clawed frogs can live a long time. Tinsley and Karen Tocque used skeletochronology and estimated that as many as 5 percent of the breeding individuals in a wild population of Couch's spadefoot toad (*Scaphiopus couchii*) in the Sonoran Desert of Arizona were more than 10 years old. Using the same technique, they estimated that some females reached 13 years old and that the maximum longevity for males was 11 years old.

The longest continuous mark-recapture field study of amphibians, including 16 species of frogs, was initiated in 1978 by the Savannah River Ecology Laboratory in South Carolina at a natural wetland known as Rainbow Bay. According to David Scott, the longest approximate time periods between first and last captures of various species of frogs have ranged from 1 to 6 years. Some species, such as ornate chorus frogs, appear to be relatively short-lived (2–3 years), whereas others (e.g., eastern spadefoot toads and southern toads) are known to live at least 6–7 years. In a mark-recapture study more than 9 years across multiple sites in Ocala National Forest, Florida, Cathryn H. Greenberg and George W. Tanner documented a 7-year-old spadefoot.

Chapter 7

Food and Feeding

What do frogs eat?

With few exceptions, all adult frogs and toads eat other animals—that is, they are carnivores. Frogs are well known for their insect-eating capabilities, and, in general, insects play a major role in their diet. However, frogs will eat any animal small enough for them to swallow, and because most frogs are small animals, insects and other small invertebrates, which are abundant in most ecosystems, form the bulk of their diet. In contrast to adult frogs, most tadpoles are herbivores, eating algae they scrape off of rocks and aquatic vegetation. Some tadpoles also feed on the detritus that accumulates at the bottom of the aquatic habitats in which they live.

Although most frogs are generalists, some frogs specialize on particular types of prey. The Mexican burrowing toad (*Rhinophrynus dorsalis*) is a dumpy, short-legged frog adapted for burrowing underground. It rarely surfaces, except to breed, and feeds exclusively on ants and termites. The snout of the Mexican burrowing toad is pointed, and its small specialized mouth can engulf large numbers of tiny insects. The crab-eating frog (*Fejervarya cancrivora*) of Southeast Asia eats mostly small crabs in its mangrove habitat.

Some large frogs have mouths big enough to eat other animals approximately their own size. The horned frogs of South America (genus *Ceratophrys*) are sit-and-wait predators capable of eating small rodents, small birds, and other frogs. The African bullfrog (*Pyxicephalus adspersus*) is similar to the horned frogs of South America but grows much larger and can eat even larger prey. Because the North American bullfrog (*Rana catesbeiana*) is large, it can eat nearly any frog with which it shares its habitat.

American bullfrogs (*Rana catesbeiana*) will eat almost any kind of animal they can overpower and get in their mouth, including large crayfish.
Courtesy Cris Hagen

The bullfrog has been introduced by humans into many areas of the world where it is not native, which has posed a problem for native frogs that have not evolved to share a habitat with another frog that can eat even the largest adults of their species.

As a rule, strict carnivores do not eat plants, but frogs may accidentally ingest plant material while struggling to swallow their prey. There are, however, a few frogs, that as adults develop the ability to incorporate significant amounts of plant material into their diet. The Indian five-fingered frog (*Euphlyctis hexadactylus*) is found in India and Bangladesh, and as it grows, the adult switches from eating insects to a diet of nearly 80 percent leaves and 20 percent mostly invertebrates. The Brazilian treefrog (*Xenohyla truncata*) is the only species of frog known to specialize on fruit. Research conducted by Hélio R. da Silva and his colleagues near Rio de Janeiro, Brazil, showed that this species consumed a variety of fruits, which formed an essential component of their diet. This treefrog did not specialize on any particular type of fruit but basically ate fruit according to its availability.

Research on the diet of anurans can provide important information for their conservation. Ann Anderson and her colleagues conducted research on three species of sympatric toads found in the High Plains of Texas and New Mexico. These toads all breed in playa wetlands, temporary lakes that form during spring or summer rains. Researchers found that all three toads depend on species of invertebrates, especially beetles, found in the grasslands surrounding the wetlands, thus emphasizing the need to conserve not only the wetland habitats where these species breed but also significant critical habitat surrounding the wetlands.

Do frogs chew their food?

Frogs primarily eat their prey whole. Frogs may chew their prey to a small degree when positioning it in their mouths for swallowing, but mostly, prey is swallowed whole and often alive. Most frogs have teeth in their upper jaw, called maxillary teeth that help them to grasp and to hold onto prey. Many frogs will use their forelimbs to force prey into their mouths and their tongues to help push food backward and down their throat. Frogs will depress their eyes into their sockets when swallowing, and it has long been thought that this pressure may further help food to move into a frog's esophagus. This contention had never been confirmed until Robert Levine of the University of Massachusetts Amherst and his colleagues tested the phenomenon in northern leopard frogs (*Rana pipiens*). They showed that, without the ability to depress their eyes, leopard frogs showed a 74 percent increase in the number of swallows required to ingest crickets. Thus, although eye depression is not necessary for swallowing, it certainly aids a frog in moving food into its stomach. Closing and depressing the eyes while consuming food also protects the eyes from harm struggling prey might cause.

The Surinam toad (*Pipa pipa*) is entirely aquatic and uses a unique method for capturing prey. This species rests in the water and waits for potential prey, primarily small fish, to swim within a few inches of its mouth. When the prey approaches too closely, the frog opens its mouth rapidly and expands the floor of its mouth and throat creating a rapid vacuum that sucks in the prey.

Why do some frogs have such long tongues?

Frogs and toads are sit-and-wait predators. That is, they pick a good spot and wait for something to come by that they can eat. Being able to respond rapidly when potential prey approaches is important, and thus, many frogs have developed mechanisms that allow them to capture prey quickly and effectively. Many frogs have elongated tongues that can be used to secure and bring prey quickly into or near the mouth, where it can be grabbed by the frog's jaws. Many species have tongues that attach to the front of their mouths that can be extended by flipping it out toward the prey. Toads have a sticky end to their tongue that sticks to insects and other small invertebrates. Some groups of frogs, such as those in the family Discoglossidae, do not have an extendable tongue. Frogs such as the African clawed frog (*Xenopus laevis*) and the Surinam toad (*Pipa pipa*) have no tongue.

The Mexican burrowing toad, *Rhinophrynus dorsalis*, has a tongue and mouth specialized for feeding on ants. Linda Trueb and Carl Gans studied

Frogs generally catch their food alive and swallow it whole without chewing, as seen with this southern toad (*Bufo terrestris*) that has captured a small earth snake (genus *Virginia*).

Courtesy Thomas Luhring

the mechanism for tongue protraction in this unusual frog and found that it uses hydrostatic pressure in conjunction with contraction of its throat muscles to protrude the tongue forward to capture its prey. They concluded that this specialization was ideally suited for capturing small insect prey (ants and termites) in restricted areas such as the subterranean burrows where these frogs live.

How do frogs find food?

Frogs exhibit various strategies that other animals use to locate food. Exactly how frogs and toads know to go to certain places where food is more abundant is unknown, but in general, their instinctive behavior results in positioning themselves within their environment in areas where they are likely to encounter a meal. Frogs and toads are ambush predators—that is, they sit and wait for an animal suitable for eating to move close to them. Where they set up their ambush spot is key to their foraging success. Most feed at night when insects are active and when they are not easily observed by potential prey or potential predators. Night activity also likely decreases evaporative water loss. Some species have learned to "hang out" near lights at night where insects are attracted in large numbers. Toads can often be found congregating under streetlights, and treefrogs are often found during the summer near or on windows where insects tend to gather. Most frogs recognize food by movement, but some species find food by smell. The marine toad (*Bufo marinus*) is known to feed on dog food directly out of dog bowls in people's backyards. It is presumed that aquatic species rely heavily on odor to locate food but likely require visual stimulation to respond to and to capture their prey. Likewise, species such as the Mexican burrowing toad (*Rhinophrynus dorsalis*) and narrowmouth toads (genus *Gastrophryne*) that spend most of their lives in underground retreats probably use chemoreception (e.g., smell) as a primary mechanism to find their food, which mostly consists of small insects, including ants and termites. Some have actually been documented living within termite nests.

Do frogs drink water?

Water is extremely important for frogs and toads. Because most frogs have relatively porous (or permeable) skin, water easily evaporates from their bodies. Frogs have various adaptations that reduce water loss, but how they actually intake water is interesting. Frogs do not drink water with their mouths like humans. Although they may obtain some water from their food, most of their body water is absorbed through their skin. Because the ionic concentration of frog body fluids is higher than surrounding water, water moves by osmosis into their skin. Terrestrial species of frogs and toads, such as the common toads in the genus *Bufo*, have highly vascularized skin on their bellies that enhances water absorption. This specialized skin, known as the *pelvic patch*, is present on their bellies and on their thighs. Remarkably, they can even absorb water from moist soil. Studies have shown that up to 80 percent of water intake of some toads takes place across this pelvic patch. In general, frogs and toads that live in drier environments, such as the spadefoots of the North American deserts, are better at absorbing water from their environments than are species such as leopard frogs found in wetter habitats.

How do desert frogs keep from drying out?

Some species of frogs have special adaptations that allow them to reduce dramatically the loss of body water across their skin. Leaf frogs (genus *Phyllomedusa*) live in the American Tropics and have developed the ability to make their skin essentially impermeable to water. During dry periods, these frogs produce lipids (i.e., fats) from special glands in their skin. The secretions are spread over their entire body with their legs, and as the secretions dry, the skin becomes shiny like wax. The wax is completely impermeable to water and, thus, allows the frog to withstand long periods of the dry season without losing body water.

Frogs in the genus *Chiromantis*, known as the waterproof frogs, live in dry areas of Africa and have skin that reduces water loss substantially. By increasing the number of skin cells known as iridiphores that reflect light, the frog is able to avoid substantial dehydration because it absorbs very little light from its surroundings, thus keeping its temperature relatively low. Some evidence suggests that the iridiphores also provide waterproofing by making the skin less permeable. Several species of frogs, such as the desert-dwelling Australian frog (*Cyclorana novaehollandiae*), can make cocoons from multiple layers of shed skin that protect them from dehydration. Dead skin cells form an impermeable barrier to water loss and can allow a frog to go for long periods in a dormant state without dehydrating. Other species that use this approach include the African bullfrog (*Pyxicephalus adspersus*), the Mexican treefrog (*Smilisca baudinii*), and the horned frogs of South America (genus *Ceratophrys*).

What do tadpoles eat?

Tadpoles primarily eat algae and other microorganisms living in water. Many have mouthparts specialized for scraping algae off rocks and other surfaces. Many use filtering mechanisms to remove organic matter from water. Some tadpoles have special adaptations on their gills that trap particles such as algae and other single-celled organisms and then direct those to their esophagus. Tadpoles have large muscles that expand and contract their throat allowing them to move substantial amounts of water into their mouths and through their gills where particles are trapped. Of course, moving water across their gills is vital for respiration as well.

Some tadpoles are known to scavenge, feeding on the bodies of dead animals. Others can even be cannibalistic, feeding on their siblings and other members of their species (see next question). Some, such as the treefrogs that lay their eggs in leaf axils, eat unfertilized eggs that the female lays periodically on visits. Janalee Caldwell and her colleagues have studied the feeding of the tadpoles of poison dart frogs (genus *Dendrobates*) and found that, unlike most other tadpoles, these species feed frequently on insect larvae found in tiny pools of water where they live. Sometimes, they are even known to eat their own kin. Caldwell concluded that such cannibalistic behavior is probably a by-product of the evolution of carnivory in these tadpoles. Tadpoles of species that have direct development from eggs (i.e., hatch as completely formed froglets) do not eat anything but obtain all their nutrients in their egg yolk.

Tadpoles of the Sri Lankan rock frog (*Nannophrys ceylonensis*) eat microscopic organisms they scrape from rocks, but their diet changes from mostly herbivorous in which they eat microflora to mostly carnivorous (microfauna) during the early to late tadpole stages. Interestingly, the length of the intestine decreases during development as well; a long intestine characterizes herbivores and a shorter one is typical of carnivores.

Why do tadpoles of some species eat their siblings?

Tadpoles of some species of frogs will sometimes eat their siblings. This phenomenon, known as *siblicide*, may seem cruel and maladaptive. That is, eating your sibling who carries similar genes may appear to be something that would be selected against. However, under certain circumstances, it may be advantageous for some individuals of certain species of frogs to eat their brothers and sisters. David Pfennig has studied this phenomenon in spadefoot toads of the American Southwest deserts and found that several species have tadpoles that sometimes develop into cannibals. Cannibals develop much larger jaws and have beak-like mouthparts used to kill and eat their siblings. Most scientists believe that environmental factors may determine whether some individuals transform into cannibals. These species often lay eggs in wetlands that may dry and in which food may be limiting. Thus, it is advantageous for the species to sacrifice some individuals so that others can survive, transform into adults, and eventually reproduce.

Chapter 8

Frogs and Humans

Do frogs make good pets?

Some species of frogs and toads make good pets, and many are attractive and interesting. Keeping a pet frog or toad can be an excellent opportunity for a child to learn to appreciate animals and to learn about the responsibilities required when caring for a pet. However, if you are looking for a pet that responds to your commands or likes to sit in your lap, we recommend a dog or a cat. Frogs are not the kind of pets that enjoy people. Most frogs would rather avoid people. That said, in captivity, some species of frogs and toads learn to tolerate people's attention, and some become accustomed to being held. Raising a pet frog or toad can be a remarkable experience for children. Raising a tadpole to adulthood requires learning about the needs of the animal and can sometimes be a challenging and rewarding experience for child and parent.

Having a pet frog or toad requires a willingness on the part of the pet owner to accept certain responsibilities. First, acquiring a pet frog should be done in a responsible manner. In some areas, all frogs or some species may be protected by law. Anyone capturing a frog to keep as a pet should know the laws governing the capture of that species. It may be best to obtain a captive-bred frog as a pet. Some species of frogs and toads do not make good pets. That is, they do not thrive in captivity and die within a short time. This may be due to the owner's inability to meet the species' particular requirements. For example, it may be difficult to obtain sufficient quantities of small insects for a narrowmouth toad, and thus, this species would likely not thrive over long-term captivity. Other species may not thrive for unknown reasons. Careful research before acquiring a pet frog can help avoid such problems.

Many species of frogs do well in captivity and require minimal care compared with mammalian pets such as dogs or hamsters. Easy-to-care-for species include those that require little maintenance, have few specific habitat requirements, eat readily, and can tolerate a moderate range of temperatures. Most frogs and toads do not like being handled, but some treefrogs will learn to sit calmly on one's finger if handled gently. Most frog and toad species are secretive and, if given the opportunity, will hide in their cage. Many temperate treefrogs make good pets as do many species of toads (genus *Bufo*). Other species often sold in pet stores or by online animal dealers that are relatively easy to care for include the South American horned frogs (genus *Ceratophrys*), White's treefrogs (*Litoria caerulea*), African bullfrogs (*Pyxicephalus adspersus*), and the attractive fire-bellied toads (genus *Bombina*).

Many people keep various species of poison dart frogs (genus *Dendrobates*) as pets. Because of their bright colors and diurnal activity, poison dart frogs make attractive displays. They lose most of their toxicity in captivity, and some species can be hardy given the right captive conditions. However, because they have fairly restrictive habitat requirements and must be fed tiny insects, which are sometimes hard to obtain, only experienced pet frog owners should keep these species, especially considering their potential toxicity. Smaller treefrogs native to the southeastern United States are often sold in pet shops and are relatively inexpensive. These include gray treefrogs (*Hyla versicolor* and *H. chrysoscelis*), green treefrogs (*H. cinerea*), and barking treefrogs (*H. gratiosa*), all of which are attractive, often brightly colored, and tend to become somewhat tame in captivity. These frogs, if purchased healthy, do well in captivity if given basic care. In general, we recommend against keeping most of the "true" frogs (family Ranidae) in captivity because they tend to leap and dash themselves against the sides of their cages.

Unfortunately, many people who buy or capture a pet frog tire of caring for it after a while. Before purchasing or capturing a frog, carefully consider your ability and motivation to care for it over time. Many species can live several years, and their longevity should be considered before deciding to keep a frog as a pet. There are few options for responsibly getting rid of a pet frog. Ideally, a frog owner can give the pet to another person interested in caring for it. Under no circumstances should a nonnative frog or toad be released into the wild. Nonnative frogs have been released in some parts of the world and have, in some cases, become invasive species by detrimentally affecting native ecosystems. Caution should be exercised even when native frogs are released to the exact location where they were captured. Captive frogs may acquire a disease such as chytrid fungus or red-leg, which is easily transmissible to wild frogs and can devastate the local population. Most scientists recommend that no frog should be released back into the wild after living in captivity as a pet.

Although most people consider frogs and toads harmless to humans, some captive frogs and toads, like many other animals, can pose substantial dangers to their owners if they are not properly cared for. Keeping frogs and toads as pets under unsanitary conditions can result in a bacterial infection known as salmonella, which is transmissible to humans. Salmonella infections have resulted in human deaths. Consequently, always carefully wash your hands with soap before and after handling any frog or toad. Many frogs and toads also have skin toxins that can irritate the eyes, nose, and mouth of humans. Careful hand washing after handling pet frogs and toads will help to prevent toxins from causing problems for pet owners. If you handle your frog or a toad, you should always do so gently, which will usually cause less stress on the animal, reducing the amount of toxin it releases. Regardless, never touch your eyes, nose, or mouth after handling an amphibian until you wash your hands thoroughly with soap. Even mildly toxic species can cause a burning sensation if not properly handled.

Where can I get a pet frog?

There are basically two ways that you can get a pet frog. You can either purchase one or capture a frog to keep as a pet. There are advantages and disadvantages to each. Impulsively buying a pet frog is never a good idea. Only after considerable research about what species make the best pets and how to care for them should you consider acquiring a pet of any species. Captive-bred frogs make the best pets and typically do much better in captivity than wild-caught animals. If you purchase a frog, make sure it is captive bred and not captured from the wild. Oftentimes, wild-caught frogs have been improperly cared for during shipment and are diseased or starved. Some pet shops may not properly care for their animals, and frogs purchased from them may be unhealthy. Carefully inspecting potential pets before purchase is critical. In addition, asking the pet store personnel questions about the animal's care and origin can be helpful in making the proper decision. Captive-bred frogs and toads can be purchased directly from breeders. Such specimens are usually much healthier and less expensive than frogs and toads purchased from local pet stores. Breeders often know much more about their animals than pet shop personnel and can provide helpful advice about selecting and caring for a frog.

Capturing a frog to keep as a pet can be fun and can be a remarkable and memorable experience. Treefrogs and toads can often be collected eas-

Bullfrogs are fun to catch but do not make good pets for children because they tend to jump against the walls of a terrarium and may injure themselves. Courtesy Mike Gibbons

ily if you know where and when to look. The breeding season, when most species are congregating around breeding sites, is generally the easiest way to find frogs and toads. Removing one or a few individuals of a large population will have minimal impact on the population. However, one should never remove animals from small populations that may be detrimentally affected by the removal of breeding adults. Collecting permits are often required to capture even common species, so take care to make sure all laws regarding taking native animals are observed. Finally, anyone that removes a frog or a toad to keep as a pet must recognize the ethical responsibility for caring for any captive animal properly.

How do you take care of a pet frog?

Many species of frogs and toads are relatively easy to care for, but some require special conditions. We will not provide all details required to properly care for all species of frogs, but we highly recommend the reader consult both books and online materials that provide valuable information about the proper maintenance of frogs and toads. Because they are ectotherms (cold-blooded), they require less food than comparably sized birds or mammals and produce less waste. A small- to medium-sized aquarium can often be converted into a terrarium that will serve well as a cage for many species. Most frogs can easily be fed various species of insects such as crickets, which can be purchased from most pet stores. Many experts recommend "dusting" crickets with calcium and vitamins so that pet frogs receive the required nutrients to maintain health and to grow. Some frogs, such as horned frogs (genus *Ceratophrys*) and African bullfrogs (*Pyxicephalus adspersus*), will consume mice, which can also be purchased at pet stores.

Frogs kept in unsanitary conditions can develop diseases contagious to other frogs. Consequently, frogs' cages should be checked and cleaned regularly. Maintaining the proper temperature is important for most frogs and toads. Fortunately, many species do well at typical room temperatures (e.g., 70–75 degrees Fahrenheit). Some frogs require higher humidity levels. Misting the cage daily with a water-filled spray bottle is an easy way to keep the humidity relatively high. Many books and Web sites provide helpful information to anyone interested in keeping pet frogs and toads. We provide the names of such resources in an appendix at the end of this book.

Are any species of frogs dangerous to humans?

Some frogs and toads are potentially dangerous to humans. However, no frog is dangerous unless it is first handled or comes into direct contact with a person. Many frogs and toads produce toxic secretions from their skin, which can endanger humans and other animals. Toxins of nearly all species can irritate eyes and mucous membranes. In some species, toxins are more potent and can conceivably cause serious illness or death if ingested. The marine toad (*Bufo marinus*), a large species native to tropical America and introduced into Florida, Hawaii, and Australia, can produce a significant amount of relatively potent toxin from its large parotoid glands. Many dogs have died from biting them and conceivably, if a person bit or chewed on one of these frogs, they could die, although we are unaware of any human deaths resulting from this species. The large Colorado River toad (*Bufo alvarius*) can also produce toxins in addition to chemicals that produce a hallucinogenic effect in humans. A good rule of thumb is to take caution not to ingest any toxins from any toad.

The most well-known toxic frogs are the poison dart frogs from Central and South America. These brightly colored species sequester toxins from ants and other insects they eat, and some, such as the golden poison frog (*Phyllobates terribilis*), have such potent skin secretions that they block transmission of neural signals and can leave important muscles, such as the heart, unable to contract. Some scientists have suggested that this species is likely the most toxic animal on earth, and it is estimated that one individual frog contains enough toxin in its skin to kill up to 20 humans. In captivity,

Horned frogs (genus *Ceratophrys*) are attractive and easy-to-care-for pets that reach body sizes large enough for them to eat mice.

Courtesy Aubrey Heupel

if they are fed a diet of small crickets or other insects, the poison dart frogs will eventually lose their toxicity. However, some species may retain some toxin for years, and thus, all members of this group should be treated with caution.

Although not deadly, some species of large frogs, such as horned frogs (genus *Ceratophrys*) from South America and the African bullfrog (*Pyxicephalus adspersus*), can inflict painful bites that sometimes break the skin. These species do not have large teeth but have large bony projections called *odontoid processes* that look and feel like large teeth. Their large mouths and powerful jaws can latch onto a finger and can be difficult to dislodge. One of us (MD) has personal experience with this and recommends against testing the biting ability of these frogs.

Do frogs feel pain?

Yes, frogs feel pain. Pain is an adaptive response to a harmful or potentially harmful stimulus that signals an animal to protect itself by avoiding that stimulus. That is, if it did not hurt when you put your hand on a hot stove you might leave it there for an extended period of time, resulting in significant injury. If frogs are injured or harmed in some way, they have the neural capabilities and receptors to perceive and respond to pain. Whether that pain resembles human sensations is difficult to gauge, but frogs react in predictable ways when presented with painful stimuli. Frogs do not usually vocalize their pain, although the distress call of some frogs may be in response to certain types of pain or fear.

Frogs: The Animal Answer Guide

Local wetlands are ideal sites to find many species of frogs that are native to a region, such as this barking treefrog (*Hyla gratiosa*).

Courtesy Thomas Luhring

Why do toads urinate on people when they are picked up?

Some frogs and toads are well known for their ability to store water in their bladder. Toads that live in the deserts can store up to 30 percent of their body weight in water in their urinary bladder. This water is critical for their survival during dry periods. When toads are captured, they will often release urine from their bladder. Some people have speculated that this reduces the weight of the toad, allowing it to jump farther and to make escaping more likely. However, such an adaptive benefit has not been demonstrated. Others believe that releasing urine from the bladder startles a would-be predator and may cause the predator to release the toad. This, too, has not been documented. For toads in dry environments, losing water in this manner may be detrimental. Every behavior exhibited by an animal does not necessarily have an adaptive purpose. As humans might urinate when frightened, toads may do the same thing. Whether this benefits humans or toads is equivocal.

How can I see frogs in the wild?

Searching for frogs and toads can be exciting, educational, and fun for people of all ages. Look for frogs and toads at night when they congregate to breed around aquatic habitats. Listening for them is frequently the best way to know they are present. Even a neighborhood pond will usually have at least a couple of species of breeding frogs. If you learn to recognize the calls of the species inhabiting the area where you live, then you can be a more effective "frog hunter." In most regions, you can hear frogs calling at nearly anytime of the year, but most call during spring and early summer. Warm, rainy nights are usually best. Some species may call during winter or early spring, and these often call during the daytime because of the cold temperatures at night. Spring and summer breeding frogs typically call at night. To search for them, triangulate a calling individual by getting a bearing on its location from several spots. Then use a flashlight to find the animal. If you are quiet and careful, you can often see the frog actually calling. Some species call from the water's edge, some call from vegetation overhanging the water, and others call while floating on the water surface. A few call from underwater. Some species are secretive and easily disturbed and can be particularly challenging to find.

Another way to find frogs is by driving country roads at night during rainy periods. To avoid desiccation, frogs will often travel during rainy nights and can often be seen from a vehicle. Of course, caution should be exercised when leaving your vehicle to capture frogs crossing the road. Although most are active at night, some species, such as cricket frogs (*Acris*), may be seen along muddy banks of wetlands and ponds during the daytime.

Chapter 9

Frog Problems (from a human viewpoint)

Why should people care about frogs?

People should care about frogs for many reasons, several of which are explained in other parts of this book. Among the basic reasons are that frogs are an important part of the world's ecosystems, serving as both predators and prey for countless wildlife species. Another reason that people should care about frogs is that they have had major roles in both ecological and medical research. Frogs have been used in ecological research and as effective bioindicators of environmental conditions, in some situations revealing the presence of pesticides, herbicides, and heavy metals that could harm humans. In one dramatic instance, at the U.S. Department of Energy's Savannah River Site, finding frogs with detectable levels of radiation in their bodies confirmed the presence of radioactive contamination in a series of holding basins. One study documented that bullfrogs, as well as both the shell and muscles of slider turtles, were radioactive. The discovery of radioactive animals led to assessments that the pond, which was near the site boundary, had radioactive sediments that could be transferred to wildlife. Bullfrogs and turtles are both local dietary items for many people in the region, and both bullfrogs and turtles were capable of traveling overland to nearby private ponds. The previously open-access holding basins were subsequently fenced in and ultimately drained and filled in.

In the medical field, bullfrogs and leopard frogs are used in basic laboratory courses in comparative anatomy and physiology and have been especially important in training doctors and nurses. Moreover, frogs have been the experimental animal used for several medical discoveries of historical significance, including bioelectricity and cloning. Frogs also have great potential practical value in medicine. Toxic chemicals produced in the skin of the poison dart frogs of Colombia offer the promise of a painkiller many times more powerful than morphine.

Many people care about frogs because some are edible. Although they are primarily popular as a delicacy rather than as a food staple, some frog species possibly can meet the guidelines required for a sustainable use product, a food source that can be produced at a sustainable level, if over-exploitation is contained. Frogs of many kinds and in many places are appreciated as aesthetic parts of our natural heritage because of their array of color patterns, musical sounds, and interesting behaviors. Unfortunately, some people may only perceive frogs as pests because of their loud night-time choruses, occasional breeding congregations followed by egg laying in ornamental fish ponds, or their visitations to home swimming pools.

Humans are responsible for managing and conserving wildlife and natural ecosystems. Frogs are important components of most natural terrestrial and freshwater systems. Thus, we have an ethical obligation to care about and conserve frogs and their habitats.

Are frogs pests?

Native frogs in their natural habitats are not pests and should not be considered as such.

Frogs do not carry transmissible diseases to humans, although improper care of a pet frog can sometimes result in salmonella poisoning. Keeping a pet frog's cage and water clean at all times usually eliminates the risk of salmonella. No frogs are venomous, although many are poisonous, ranging from mildly poisonous to potentially deadly. Typically, frogs are harmful only when someone picks one up and the frog's skin contacts sensitive tissues of the eyes, nose, or mouth or the poison gets into the bloodstream through a cut or abrasion.

A common complaint about frogs is the noises they make, especially during breeding periods. A large chorus of spadefoot toads and treefrogs can be deafening. However, when such sounds are made by native frogs, they should be viewed as a wildlife experience to marvel at while it lasts. Most situations in which frogs call so loudly that a person nearby cannot be heard last only one or two nights, and then the number of calling males decreases until another suitable rain. Although most people would appreciate the opportunity to experience such an event, we have received complaints from people about native frogs calling near their homes. When the chorusing of native frogs is viewed as a major problem, perhaps they should consider living in areas without frogs, such as Antarctica, Greenland, or Iceland. Probably everyone would be happier, including the frogs. Com-

Among the most notorious frog pests in the world is the cane toad (*Bufo marinus*), a native of tropical America. A major negative impact from cane toads has been in Australia where the species was introduced in the previous century to control beetles that were pests on crops, but the toads are toxic to native predators and are voracious predators themselves on native frogs. Courtesy John D. Willson

plaints about noisy frogs have arisen in some areas of Hawaii where high-pitched calls of nonnative coqui introduced from Puerto Rico have become highly unpopular, and efforts are under way to control this invasive species, including spraying them with citric acid or with hot water.

Homeowners with swimming pools may consider frogs as pests because they have to be removed from skimmers and may even lay eggs in the pool. This is usually a short-term and infrequent problem. One should consider it a positive opportunity to observe native frogs in the neighborhood. If pool water is treated with chlorine and other chemicals, frog eggs probably will not hatch. Frogs crossing highways in large numbers during rainy periods in the breeding season can be run over by cars. Of course, this is a bigger problem for frogs than for humans, but no normal person enjoys running over any kind of wildlife and a large toad or bullfrog beneath a tire is not a sound to go unnoticed.

Viewed objectively, humans are the pests and create situations that make

Frog Problems (from a human viewpoint)

frogs appear to be the problem. It is difficult to fault invasive frog species that negatively affect native frogs and other animals because, in most cases, humans brought them there.

Do frogs have diseases and are they contagious to humans?

Frogs in the wild rarely, if ever, have a disease that is contagious to humans. However, frogs are susceptible to many diseases, some of which can be lethal and which, in some instances, have led to the extinction of certain species. Three serious and widespread diseases of frogs are a fungus, a type of bacteria, and a virus.

The chytrid fungus, or *Bd* (*Batrachochytrium dendrobatidis*), is implicated in major amphibian declines in all parts of the world. Amphibian biologists have offered opinions and data regarding the origin of the disease, how it is spread, its virulence, and even why some species of frogs are more susceptible than others. Not all scientists agree with all details, although most agree that chytrid fungus has devastated frogs in many regions. However, frogs infected with chytrid fungus are not contagious to humans.

Red-leg, or red-sore, disease is most commonly seen in captive frogs (and in unhealthy fish), sometimes occurring a short period after capture. Red-leg is most apparent on the underside of a frog as hemorrhaging and is caused by a bacterium (*Aeromonas hydrophila*). Red-leg is contagious to other frogs but not typically contagious to humans, except under unusual circumstances, such as a digestive system complication in humans with impaired immune systems.

Like other animals, frogs are susceptible to various viruses, but one group in particular, known as ranaviruses, has been identified as a source of mortality in frogs in the wild, as pets, and in laboratory environments. Although most ranavirus research on frogs has focused on species that inhabit North America, ranaviruses have been reported in frogs on most continents. Research must still be performed to understand methods of transmission and differential responses among frog species and localities, but humans are not believed to be susceptible to any of the viruses that affect frogs.

Although parasitic trematode flatworms technically are not a disease, they have been implicated in exacerbating developmental malformations in frogs infected with the parasites and whose immune responses have been compromised by agricultural pesticides. As with research on frog diseases, further studies are needed to understand the association between frog body malformations, trematodes, and agricultural contamination. Humans do not host trematode parasites that infect frogs but considering whether toxic contaminants in an area might be problematic for humans is advisable.

Ironically, one ailment that humans can contract from frogs, even if the frog is unaffected, is salmonella poisoning. Salmonellosis, a bacterial infection that causes intestinal complications, can range from mild food poisoning to lethal in some instances. Sources of salmonella poisoning include contaminated foods such as eggs or chicken as well as the cages and water bowls of many pets, including pet frogs. In 2009, the Centers for Disease Control and Prevention (CDC) issued a warning that African dwarf clawed frogs (genus *Hymenochirus*) sold as pets were responsible for salmonella cases in parts of the United States. According to the CDC, most victims were children who owned pet African dwarf clawed frogs from a single commercial pet distributor. Contact with dirty water from the frog's container to the pet owner's hands is believed to have been the source of the salmonella transmission. Although salmonella is sometimes associated in this way with pet amphibians and reptiles, the frog, turtle, or snake is not really the problem; the problem is the pet owner or distributor allowing unsanitary conditions to develop.

When considering the susceptibility of humans to diseases prevalent in other animals, the human immune system remains poorly understood and in rare instances a fungal, bacterial, or viral infection from a frog could be contracted. We recommend caution when handling any frog or toad that appears sickly and that highly sanitary conditions be maintained for all pet frogs and toads. Washing hands thoroughly before and after handling any frog, even if it appears perfectly healthy, is a good practice.

Is it safe to eat frogs?

Eating frog legs is safe as long as the frogs belong to one of the large, edible species, such as the bullfrog or the edible frog (*Rana esculenta*) of Europe, and care is taken if the skin is relatively toxic, such as in the smoky jungle frog (*Leptodactylus pentadactylus*) of tropical America. Toads and treefrogs are not considered human food because of their relatively small legs and toxic skin. A frog's hind legs are usually the only edible body part, and the meat is white. The frog's skin is not eaten. Frogs "taste like chicken," but some amphibian biologists say chickens taste like frogs. A simple way to prepare frog legs is fried or sautéed. An outstanding cookbook, *The Culinary Herpetologist* (2005) by the late Ernie Liner of Louisiana, has several recipes for frog legs and for preparing other edible species of amphibians and reptiles.

Ernest A. Liner

The culinary Herpetologist

Among the frog recipes included in *The Culinary Herpetologist* by Ernest A. Liner is "Roasted Poison Dart Frog" (Campa Indian style). Courtesy Breck Bartholomew

Are frogs raised by humans for food?

As the Missouri Department of Conservation stated in its Pond Management Series, "Successful frog farming is definitely more myth than reality." Although the target audience was in Missouri and the farmed frogs referred to are bullfrogs, this applies worldwide for other species and other regions: commercial frog farming is generally not considered to be a self-sustaining occupation. Most scrutinized records of commercial frog legs sold to restaurants and grocery stores have revealed that most frogs are wild caught.

Factors operating against frog farming as a profitable enterprise are the increased chance of contagious diseases at high densities, the cost of land and the necessary water to raise tadpoles, and the cost of containment to prevent predation and dispersal. Frogs are carnivores and would need to be fed animal matter, as local insect populations would not be adequate to support a large, high-density population of bullfrogs. Only a small portion of the frog (the hind legs) is edible, so the cost of feed per pound would be prohibitive.

Frogs: The Animal Answer Guide

Madagascar bright-eyed frog,
Boophis madagascariensis,
Madagascar. Courtesy Leslie Ruyle

Map treefrog, *Hypsiboas geographicus*, Peru. Courtesy Chris Gillette

Napo lime treefrog,
Sphaenorhynchus carneus, Peru.

Courtesy Chris Gillette

Perez's snouted frog, *Edalorhina perezi*, Peru. Courtesy Chris Gillette

Pine barrens treefrog, *Hyla andersonii*, Florida. Courtesy Aubrey M. Heupel

Tomato frog, *Dyscophus antongilii*, Madagascar. Courtesy Dante Fenolio

**Polka-dot treefrog, *Hyla punctata*,
Peru.** Courtesy Chris Gillette

Queensland barred frog, *Mixophyes fleayi*, Australia. Courtesy H. B. Hines DERM

Red-eyed treefrog, *Agalychnis callidryas*, Tortuguero, Costa Rica. Courtesy John D. Willson

Smoky jungle frog, *Leptodactylus pentadactylus*, Tortuguero, Costa Rica. Courtesy John D. Willson

Sonoroan green toad, *Bufo retiformis*, Pima County, Arizona.
Courtesy John D. Willson

outh American common toad, *Bufo margaritifer*, Peru. Courtesy Chris Gillette

Southern toad, *Bufo terrestris*, **South Florida.** Courtesy Chris Gillette

Strawberry poison frog, *Dendrobates pumilio*, **Central America.** Courtesy Pierson Hill

Can a person get high from licking or smoking a toad?

Whether the activities of toad licking and toad smoking have been engaged in to any great extent remains shrouded in uncertainty, although the occasional adventure seeker has tried it. The premise is that licking the large, poisonous parotoid glands of Colorado River toads (*Bufo alvarius*) or grinding up a dried carcass and smoking it provides a hallucinogenic reaction. An account of this unusual activity by Cris Hagen, a scientist and educational specialist at the Savannah River Ecology Laboratory, follows: "Ingesting the toxins of *Bufo americanus* and *Bufo marinus* made me sick for a couple of days. I felt poisoned and wanted to throw up but never did. Smoking dried *Bufo alvarius* poison caused immediate and sometimes strong hallucinations that lasted for 10–15 minutes. It had a sharp 'poisony' taste." Because Cris attempted this experiment as a teenager and we are now colleagues of his, we are relatively certain that Cris has moved on to other forms of entertainment.

Despite the appeal of hallucinogenic drugs to some people, toxins associated with toads are complex alkaloids that can be potentially dangerous if ingested or inhaled. Several variables undoubtedly contribute to the effects, including a person's physiological state, the species of toad, and the individual toad's biological condition. Some effects could be hallucinogenic and many could be detrimental in other ways. We do not advise licking frogs or toads.

Can toads cause warts in humans?

The simple answer is, "No." Toads look warty, but their warts are not similar to warts people contract. Some warts are glands of various sizes and shapes and result from thick areas of skin that form a tough, fibrous material known as keratin. Warts in humans are caused by the human papillomavirus, or HPV (not characteristic of frogs or toads), which is contagious from one person to another. Although many people appear not to be susceptible to warts, it is possible to be infected by using towels, utensils, or other objects after they have been used by a person with warts.

How can frogs be used in research?

In addition to their value in ecological studies as biomonitors and indicator species, as a key taxonomic group, frogs have been invaluable for medical research and in addressing general principles of physiology and anatomy of vertebrates. Consequently, frogs have been used to pave the way for research, as well as for medical applications, through their use in

The poison from the large parotoid glands of adult Colorado River toads (*Bufo alvarius*) has been reported to produce a hallucinogenic reaction when smoked or tasted. The activity is not recommended for health and safety reasons, and because the toads do not appreciate it. Although herpetologist Steven Price (pictured here) is known for eating strange things at times, he is not actually licking the toad. Courtesy John D. Willson

teaching laboratories, where, for decades, preserved frogs have been a primary target for dissection. Among the notable research for which frogs have been critical study animals has been the use of frog skin to study ion transfer across membranes. In the 1700s, the Italian Luigi Galvani conducted research on frogs in the laboratory and discovered the link between electricity and the nervous system. In the 1950s, Robert Briggs and Thomas J. King cloned a frog by using a somatic cell nuclear transfer. In the 1990s, the same technique was used to create the famous "Dolly the Sheep," the first mammal to be cloned, but the experiment on the frog was the first time successful nuclear transplantation had been accomplished on a multicelled animal of any kind.

In the early twentieth century, leopard frogs (*Rana pipiens*) and the African clawed frog (*Xenopus laevis*) were used for pregnancy tests. *Xenopus* is also used for research in developmental biology and for understanding the complicated changes that happen to frogs during metamorphosis. Barbara Lom and her students at Davidson College use *Xenopus* as a model organism to study factors that affect the development of neurons between the eye and the brain. In a true bit of irony, *Xenopus* has perhaps spawned more ecological research than it did in the medical fields by being indicted by amphibian biologists as a primary carrier of chytrid fungus. Chytrid fungus is believed by some to have been broadly spread as a consequence of the accidental or intentional release of infected *Xenopus* into nonnative habitats in North America, Europe, and elsewhere. Research continues to be conducted throughout the world to determine the impacts that chytrid fungus will have on amphibians.

Frogs: The Animal Answer Guide

Chapter 10

Human Problems (from a frog's viewpoint)

Are any frogs endangered?

Frogs may be the most imperiled of all vertebrate animals. Many of the world's frogs are endangered, but "endangered" has different meanings based on the context in which it is used. Formal and often legal classifications and designations of particular frogs and toads as endangered are through several recognized and official entities. Foremost are the IUCN (International Union for Conservation of Nature), the U.S. Endangered Species Act, certain states, and some countries. Designating endangered status varies but is based on the documented biology and perceived conservation status of the particular frog species, and the process is always influenced by regional, country, or international politics. The overall concept is a good one and is effective and properly done in many cases, but the result can often be that a species is not listed when it should be, whereas another species is listed when it should not be. The latter situation is rare but has occurred. Part of the complication is that even when studies have been conducted and data analyzed and published, amphibian experts do not always agree on their interpretation and what the proper status and conservation designation of a species should be.

IUCN categorizes the level of vulnerability and the formal listing of species on an international scale by producing the IUCN Red List classification. IUCN recognizes several levels of endangerment for species and places species deemed to be in trouble into categories. The IUCN classification listings, in order of increasing peril for a species, are vulnerable (VU), endangered (EN), critically endangered (CR), extinct in the wild (EW), and extinct (E; not just in the wild but in zoo, laboratory, and

The Houston toad (*Bufo houston-ensis*) of Texas is a federally endangered species. Courtesy Dante Fenolio

personal collections as well). The categories get progressively worse. *Data deficient* means that insufficient information is available to determine the conservation status and classification of the species. Data deficient does not mean that a species is not endangered at some level; it means that no data are available to provide unequivocal evidence, which applies to frogs in most regions where they occur. A dramatic reinforcement that species can be in peril without being on the IUCN Red List of Threatened Species is the experience of Joe Mendelson, curator of herpetology at Zoo Atlanta. As an amphibian expert, Mendelson formally described more than 40 new species of frogs, toads, and salamanders between 1994 and 2010. Of those described, at least three species are believed to have already become extinct, and a dozen are near extinction. None of these species were on the IUCN Red List before 1994, and some were not on the list before their extinction.

An assessment of the global decline of amphibians was conducted by Simon N. Stuart, a member of the IUCN Species Survival Commission, and colleagues. They concluded that the declines were real and more rapid than the decline of birds or mammals. They determined that the declines were most prevalent in species living in mountain habitats and associated with streams in the American Tropics.

The IUCN Red List is dynamic because changes are made as new information is provided. The following were the listings for frogs and toads of the world at the beginning of 2010: vulnerable (561), endangered (652), critically endangered (406), extinct in the wild (2), and extinct (35). In total, 1,656 frogs and toads worldwide are vulnerable or in worse shape, a disquieting number. More than 1,400 species of frogs are listed as data deficient.

The U.S. Fish and Wildlife Service in accordance with the Endangered Species Act manages a separate threatened and endangered species list.

The Puerto Rican crested toad (*Peltophryne lemur*) is listed as Threatened on the U.S. Endangered Species List by U.S. Fish and Wildlife Service. Courtesy Dante Fenolio

The list included 10 species of frogs at the beginning of 2010. Of these, 5 were classified as threatened (T) and 5 as endangered (E). They are as follows: golden coqui (T; *Eleutherodactylus jasperi*), California red-legged frog (T; *Rana draytonii*), Chiricahua leopard frog (T; *Rana chiricahuensis*), Mississippi gopher frog (E; *Rana capito sevosa*), mountain yellow-legged frog (E; *Rana muscosa*), guajon (T; *Eleutherodactylus cooki*), arroyo southwestern toad (E; *Bufo californicus=microscaphus*), Houston toad (E; *Bufo houstonensis*), Puerto Rican crested toad (T; *Peltophryne lemur*), and Wyoming toad (E; *Bufo baxteri=hemiophrys*). Similar to the IUCN Red List, the species on the U.S. Endangered Species List change on the basis of new biological information, and the list is influenced by national, state, and regional politics.

From a conservation perspective, species become endangered for many reasons, which are discussed in-depth throughout this book. Some contributing factors include habitat destruction from urban development, road construction, deforestation, and agriculture; diseases that affect frogs; pollution, such as agricultural pesticide and herbicide runoff (especially from atrazine) and power reactor flyash; changes in weather patterns in response to climate change; and introduced species that prey on frogs.

Why are frog populations declining?

Frog populations in most parts of the world where frogs occur are declining and even going extinct at alarming rates to amphibian biologists. Anyone interested in these fascinating creatures or the health of their environment should be concerned. Various causes of declines include disease, habitat destruction, pollution, invasive species, and other factors discussed in this book.

Frog species in highly restricted ranges are often more susceptible to declines than frogs in widespread ranges, in part because they consist of only one or a few populations so that the decline of a single population represents a loss to a greater portion of the species. Also, species with re-

Potential reasons for the dramatic population declines of the critically endangered lemur leaf frog (*Hylomantis lemur*) of Costa Rica and Panama include habitat loss and chytridiomycosis. Courtesy Dante Fenolio

stricted ranges often live in distinctive habitats with special environmental features that have led to specialized traits in the species. Thus, the species may not be adapted to adjust to changes in the habitat, many of which are caused by human activities.

High montane species that rely on cool, moist conditions may be more susceptible if drier, warmer conditions prevail. The presumed extinction of the golden toad (*Bufo periglenes*), which was endemic to a few mountaintops in Costa Rica, was attributed, at least in part, to changes in global climate that led to weather changes in the Monte Verde locality where the frogs occurred. Another excellent example is Rose's ghost frog (*Heleophryne rosei*) of South Africa, which is confined in its natural habitat to about three square miles. Adults have strong, webbed hind feet for swimming in fast-moving streams, and tadpoles have suctorial mouthparts for clinging to rocks in the stream while grazing on algae. Because Rose's ghost frog depends on moving water, activities that change the character of the stream can be detrimental to the species. The building of upstream dams that have reduced water flow threaten the continued existence of the stream habitat essential for the species to persist. Unless regulations are established and enforced to maintain and restore its stream habitats, Rose's ghost frog likely faces extinction.

Species with broad geographic ranges can also suffer reductions in numbers and sizes of populations from unnatural causes. Although chytrid fungus (*Batrachochytrium dendrobatidis*, or *Bd*) is a naturally occurring organism believed to have originated in Africa, the disease has apparently spread to the United States, Australia, and the Tropics of Central America. One suspect in the spread of chytrid fungus in some areas is the release of African clawed frogs (*Xenopus*) into nonnative habitats where they become established in the wild, including in California and England. Un-

Frogs: The Animal Answer Guide

fortunately, *Xenopus* can carry chytrid fungus, which can be passed on to native frogs, leading to their demise. Although all of the factors involved in the widespread distribution of chytrid fungus and the differing susceptibility among frog species are not firmly established, the disease and its human assistance in reaching some habitats have clearly caused the decline of many frog populations.

Individual frogs infected with a disease, such as chytrid fungus, ranaviruses, or *Aeromonas* (see in-depth descriptions later in this section), may find the effects of the diseases are exacerbated directly and indirectly by environmental factors caused by humans, including pollution from pesticides, climate change resulting in changes in regional weather patterns, increased and abnormal levels of ultraviolet radiation, and deforestation.

In 1999, Ross A. Alford and Stephen J. Richards wrote a thoughtful review of the global decline of amphibians and discussed the issues involved in scientifically assessing the extent and severity of the problem. They observed that even under natural conditions most populations of frogs are more likely to decline than increase. This phenomenon can be readily illustrated with a study of bullfrogs we were involved in at Ellenton Bay, a freshwater wetland in South Carolina in 2003–4. During the 5 years before the study, the region underwent drought conditions: Ellenton Bay dried up, and no frogs bred successfully. Hence, the local population had declined for 5 years in a row. In 2003, an El Niño year, Ellenton Bay experienced heavy rainfall for months, the wetland filled with water, and the few remaining frogs bred, laying hundreds of thousands of eggs, many of which survived to become tadpoles, juveniles, and adults. In 2003, the population increased dramatically, making up for the previous years of decline. Similar patterns can occur in many other species of frogs in which breeding does not occur every year, even under natural conditions. However, the decline of frog populations at abnormal levels is real, and humans can probably be held responsible.

Why are frogs disappearing from some parts of the world but not others?

The disappearance of frogs in a region results from at least two important factors: (1) that frog populations are actually contracting in size or that the number of populations is declining, and (2) that an amphibian biologist or other qualified scientist has documented the declines and communicated the findings to the scientific community. Part of our perception of declines is based on credible scientific publications followed by media presentations of the findings, some of which are not always handled responsibly.

Nonetheless, frog population declines and species extinctions are definitely higher in some areas than in others. As a general rule, more species are likely to decline in areas with high numbers of species, especially when some have localized geographic ranges or specific habitat requirements, simply because more species are available to be affected. Thus, amphibian biologists have confirmed numerous species declines and the complete extirpation of some frog populations in the high-species-diversity habitat of tropical rain forests in Costa Rica. Declines are believed to be primarily a consequence of chytrid fungus and in part due to the highly specialized behaviors, dependence of species on special features of the habitat, and their restricted geographic ranges.

For unexplained reasons, frogs in some areas are more greatly affected than others. Some seemingly pristine areas have lost many of their frogs, mostly to disease that may not be related to human activity. Nonetheless, scientists speculate about why one species might be more susceptible than another. One assumption is that generalist species, such as the American bullfrog, with broad habitat requirements that range over large geographic areas tend to do fine. Bullfrogs are known to carry chytrid fungus but are not usually detrimentally affected by it. Another reason is that certain regions may have experienced the introduction of nonnative species that have become problems for frogs. Examples include *Xenopus* in California that may have been responsible for carrying chytrid fungus that infected other frogs and cane toads in Australia that eat or compete with native frogs.

Are frogs affected by climate change?

Global climate change always has and will continue to affect all sorts of wildlife in myriad ways. For some species, their geographic range will shrink or expand. Some will move to higher elevations as the lower elevations become too hot or too dry, and some will disappear entirely. Frogs will certainly be affected and, according to some determinations (as with the golden toad mentioned earlier), are likely to have already been affected.

Climate change has particularly high impact on frog species that rely on seasonal patterns of rainfall and temperature for reproduction and tadpoles that develop in aquatic habitats created by rainfall. The ornate chorus frog (*Pseudacris ornata*) of the southeastern United States breeds during winter. No one is certain what cues ornate chorus frogs to breed, but presumably cold temperatures and cold rains are critical, or perhaps short spells of warm weather following cold weather, which indicates the end of winter. If the seasonal rain patterns shift and certain regions in the Southeast go from fall to spring, with no appreciable winter precipitation, what will happen to this species? Documentation of the potential effect of climate change on amphibian breeding patterns and seasonal timing was given by Brian Todd

The ornate chorus frog (*Pseudacris ornata*) of the southeastern United States breeds during cold, rainy periods in winter. If seasonal rain patterns shift so that certain regions go from fall to spring, with no appreciable winter, what will happen to such species? Courtesy Thomas Luhring

of the University of California, Davis, and colleagues. The study was based on data collected at a single freshwater wetland (Rainbow Bay on the Savannah River Site) over a 30-year period by Ray Semlitsch, Janalee Caldwell, Joe Pechmann, David Scott, and colleagues. The study showed that most of the frog species at the site were not appreciably affected in their seasonal breeding patterns at the site but that ornate chorus frogs were breeding as much as 8 weeks earlier at the end of the 30 years than in the earlier years. Regional warming was estimated to be more than 2 degrees Fahrenheit. The unusual shift in the timing of breeding by the ornate chorus frog demonstrates the complexity of species responses to climate change.

What will be the effects on aquatic breeding species, such as most frogs in the Southeast, if droughts become more severe and more prolonged? Presumably, they will have to respond in some way as environmental conditions change. Finally, the environmental cascade effect could be a far-reaching impact of global warming on frogs. Even though a particular species might be able to adjust to shifting environmental temperatures and habitat changes, they could be affected in at least two other ways. A predator's, such as a frog's, livelihood depends on its prey. If the prey species is affected by climate change and declines, so will the predator. A second factor is that although one species of frog is not directly affected by climate, other species of frogs with which they formerly had inconsequential interactions may shift the timing of reproduction and assume new roles by becoming competitors.

Finally, the relationship between climate change and human-caused factors, such as toxic chemical pollution, habitat alteration, or even acid rain may be complex and may increase the detrimental effects of other factors that frogs have to face naturally, such as parasites and disease.

Although frogs clearly might respond to certain aspects of climate change that could have immediate or long-term impacts on frog populations, a more puzzling phenomenon is the reported response of toads at a site in Italy in 2009 to an earthquake before it happened. According to the scientific article in the *Journal of Zoology*, 96 percent of male common toads (*Bufo bufo*) left a breeding site unexpectedly, five days before a major earthquake occurred 46 miles away. The investigators considered a variety of pre-seismic cues that the toads might have responded to, including pertubations in the ionosphere, increased electrical activity, and changes in the earth's magnetic field. Perhaps the question should be, why would toads be concerned about an earthquake in the first place? They do not live in skyscrapers or use tall bridges that might collapse. Animals that flee events that are not real problems for them generally do not persist on the evolutionary scale, and we are skeptical that toads would even care whether an earthquake occurred in the first place, especially nearly 50 miles away because it would have minimal or no effect on them. The 4 percent of males that remained at the wetland and successfully bred with the females would have had a much higher reproductive success rate. In conclusion, we do not believe that toads are any more reliable as indicators of upcoming earthquakes than they are of the outcomes of sporting events or the stock market.

Are frogs affected by pollution?

Direct and indirect pollution greatly affect frogs, as mentioned elsewhere in this book, although some have challenged documentation offered by investigators about its effects. Some amphibian biologists have reported that the widespread use of the herbicide atrazine has been highly detrimental to frogs by affecting reproduction, survivorship, and resulting in the feminization of male frogs. Studies to examine the effect of atrazine have resulted in controversy because other biologists have reported that they have found no complications in frogs from using atrazine. The finger-pointing among scientists has resulted in allegations by some of those who have reported serious effects on frog populations that others—those who have not found any effects but whose research was funded by the commercial producers of atrazine—were influenced by the relationship between the funding source and the researchers. If atrazine or any other commercially produced chemical affects the well-being of frogs as indicated, the public should be informed that the product could potentially affect humans.

The effects of contaminants (such as trace amounts of arsenic, chromium, and other heavy metals) on tadpoles and adults living in aquatic systems polluted with coal ash from a power reactor have been examined by Christopher L. Rowe, Justin Congdon, and Bill Hopkins. Their findings indicated that bullfrog tadpoles from polluted habitats had higher metabolic rates when compared with bullfrog tadpoles living in natural aquatic sites. Investigators also reported that tadpoles living in ash basins were more likely to have malformed mouthparts. Such problems at a specific location may cause that population to become a "sink" population—that is, breeding occurs at this site, but because of the pollution, little if any recruitment of individuals into the adult population occurs. The population is only maintained by emigration of individuals from other healthy "source" populations.

Additional studies demonstrating that heavy metals and pesticides can potentially cause detrimental consequences in frogs were carried out in Austria by B. Grillitsch and A. Chovanec on three species (the common toad, *Bufo bufo*; agile frog, *Rana dalmatina*; and marsh frog, *Rana ridibunda*). Although specific effects were not reported, researchers indicated that their findings of high concentrations of cadmium, copper, lead, and zinc, as well as an agricultural insecticide called lindane, demonstrated the importance of nonpoint source inputs of toxic materials from wind and rain. These studies are ideal examples of how frogs can be used as a sentinel species, or bioindicators, of environmental problems that could ultimately lead to effects on humans.

Are diseases causing frog populations to disappear?

In the late 1980s and early 1990s, we became aware of something scientists had been aware of for several years—many frogs and toads were disappearing. Accounts of declines or disappearances of frog and toad populations from various parts of the world were alarming and raised concerns among scientists and the public. Although anthropogenic factors such as development and pollution were clear causes of declines, the exact causes were unknown. Declines also occurred in protected, nearly pristine habitats such as Rocky Mountain National Park and the cloud forests of Costa Rica. The potential cause of such enigmatic declines was unknown for many years, but there were numerous instances of scientists returning to a study site where they had worked for years to find scores of dead or dying frogs or, sometimes, no frogs at all. Various hypotheses were posed and disregarded as possible reasons for such declines in widely separated parts of the world. That disease was a major reason many declines surfaced in the late 1990s.

As it became clear that disease was a primary cause of amphibian declines, it was still unclear what disease or diseases were causing the problems. Several diseases known to affect amphibians were found to cause declines. Cynthia Carey and David Bradford both conducted studies in the 1990s and discovered that a bacterial pathogen known as *Aeromonas hydrophila*, which causes red-leg in frogs, was the primary source for some die-offs in the Sierra Nevada Mountains of California and in boreal toads (*Bufo boreas*) in other parts of the western United States. Andrew Blaustein of Oregon State University and his colleagues found that a fungus known as water mold (*Saprolegnia* sp.) caused major die-offs in the eggs of several frog species. Working with Blaustein, Joseph Keisecker found that climate change affected the depths at which eggs were laid, thus increasing their exposure to ultraviolet B radiation, which in turn increased the detrimental effects of *Saprolegnia* on amphibian eggs. In the late 1990s, several researchers found that viral pathogens known as ranaviruses were responsible for disease outbreaks and declines in both captive and wild populations of frogs and toads.

Over time, it has become clear that one widespread pathogen known as chytrid fungus, or *Bd* (*Batrachochytrium dendrobatidis*), has been responsible for the declines and extinction of numerous species of amphibians, especially in Australia, the western United States, and the Central American Highlands. Fungi in this group are typically found only in the soil and were not known to cause disease in any animal until it was discovered in amphibians. Karen Lips has been one of the primary researchers examining the impacts of chytrid fungus in the highland Tropics of Central America. Lips and her colleagues evaluated the patterns of *Bd*-related declines throughout the highlands of Central America and South America and concluded that a wave of *Bd* declines had proceeded southward through the Central American Highlands beginning in 1987. They also found that patterns of *Bd* declines in South America indicated that multiple introductions of the disease were supported. They found no link between climate change and *Bd*-related declines as some previous researchers had suggested.

Ample evidence indicates that *Bd* occurs in some amphibian populations where it has little impact. Questions still remain about whether *Bd* has occurred in amphibian populations for a long time and something recently has caused it to become pathogenic or whether it has been recently introduced into amphibian populations that have no resistance to its effects. Either hypothesis implies anthropogenic factors contributed to the relatively recent declines in amphibian populations related to *Bd* infections.

Although scientists knew that *Bd* caused major amphibian declines, the exact pathology of the disease was unknown for some time. Because it infected the skin of frogs and toads and because the skin of frogs is important in ion and water balance, most scientists suspected that deaths occurred because of frogs' inability to regulate water and ions effectively. In 2007, Jamie Voyles (James Cook University) and her colleagues showed that ionic imbalance detected in frogs' blood explained how chytrid fungus killed its hosts. In 2009, Lee Berger and other Australian researchers showed that such ionic disruption affected the functioning of frog muscles, namely, the heart, thus causing eventual death.

Why do some frogs have extra legs?

In 1996, some schoolchildren in Minnesota were on a field trip to a local pond. During their trip, they found dozens of malformed frogs, including frogs with missing limbs, extra limbs, underdeveloped limbs, and missing eyes. Because this discovery occurred when worldwide alarms were already high regarding amphibian declines, it garnered a significant amount of press and raised concern throughout the United States. Basically, the story was that something was causing developmental defects in frogs, and that same "something" could cause problems in humans. In many ways, the amphibian malformations received more attention than the worldwide amphibian declines and extinctions—perhaps because of the shocking images in the media of grossly deformed frogs.

Because of the attention this discovery received, reports of frog malformations came in from other parts of the United States and other parts of the world, suggesting a global problem. Some frogs undergoing metamorphosis will not develop correctly and will exhibit malformations like those found by the Minnesota schoolchildren. High numbers of frog malformations in a confined area raised concern, however. Something had to be causing these gross deformities.

Eventually, scientists discovered several interrelated factors responsible for the malformations and recognized that different factors may cause the malformations reported in other regions. Stan Sessions and his colleagues showed that a type of small flatworm known as a trematode infected tadpoles and, under the right conditions, caused many of the reported malformations. Further research eventually clarified that pesticides may increase trematode infections because of their positive effect on snail populations, one of the worm's hosts. Thus, indirectly increasing snail populations resulted in high incidences of frog malformations. Interactions among the trematodes and pollution are complex, and scientists are still working on

Limb malformations have been observed in frogs and toads throughout the world and can sometimes be traced to local pollution or parasitism within a population. The cause of the malformed forelimbs of this Woodhouse's toad (*Bufo woodhousii*) is unknown, but the abnormality did not prevent the individual from reaching adulthood. Courtesy Mike Dorcas

exactly how the interactions cause malformations and population declines. Other factors that likely play a role include the species of frog, habitat, and other human-related issues (roads, etc.). Not all studies conducted on frog malformations associated with agricultural pollution have yielded adverse results. A study conducted on the common frog (*Rana temporaria*) in Finland did not find any higher incidence of limb malformations in frogs from areas affected by agrochemicals than in natural areas in the country. Investigators suggested that environmentally friendly Finnish agricultural practices caused the lack of morphological abnormalities.

Mike Lannoo's *Malformed Frogs: The Collapse of Aquatic Ecosystems* (2008) provides a detailed history of malformed frogs and a complete summary of the science that led to a better understanding of the phenomenon and its importance. A Web site operated by the National Biological Information Infrastructure of the U.S. Geological Survey, in collaboration with the Savannah River Ecology Laboratory, provides background on amphibian malformations and the ability for anyone to submit reports of malformed amphibians (http://frogweb.nbii.gov/narcam).

Frogs: The Animal Answer Guide

How do roads affect frogs?

Throughout the world, roads have major impacts on animal populations. In the United States, there are more than 5 million miles of paved roads. Such dominant habitat alteration impacts wildlife, including frogs, in many ways. Perhaps the most visible way that roads affect frogs is mortality resulting from collisions with automobiles. Roads often pass through habitats amphibians frequent, such as those near wetlands where frogs and toads breed. Thousands of frogs can be hit by cars while migrating to breeding sites each night. The impacts of road mortality on frog populations are unknown for most areas but have been shown to reduce populations substantially. A study by a team from Carleton University in Canada headed by Lenore Fahrig studied frog populations in areas of different traffic intensity and found that (1) the number of dead and live frogs and toads per mile decreased as traffic intensity increased; (2) the proportion of frogs and toads that were dead increased as traffic increased; and (3) frog chorus intensity (i.e., calling) decreased with increasing traffic intensity. Another study in Denmark showed that road traffic could result in the decimation of 10 percent of the total population of several frog and salamander populations. Clearly, such annual reductions of the population are not sustainable long term.

Numerous efforts have been implemented to reduce the effects of road mortality on frog populations. During frog breeding season in Guelph, Ontario, local law enforcement actually closes a road at night on which many frogs traverse while traveling to their breeding site. In most places, this is not practical. Many people have advocated using tunnels under roads within which amphibians can travel safely to the other side. Such tunnels are associated with fences or walls that prevent frogs from going onto the road and direct them to the tunnels. Several studies have been conducted to determine what type of tunnel frogs are more likely to use, and some tunnels have had positive results in getting toads safely across, or under, the road. The best method to avoid road mortality in frogs and toads is to examine the habitat before roads are built, making sure areas of high amphibian traffic are avoided.

Perhaps more important are the indirect effects of roads on frog populations. Construction of roads often requires destruction and alteration of considerable habitat on which many species of wildlife, including frogs, depend. In the United States, destruction of wetlands for road construction must often be mitigated by building or preserving wetlands at other locations. Another problem is runoff from roads. Runoff can result in pollution known as *nonpoint source* pollution. Such runoff can contain many contami-

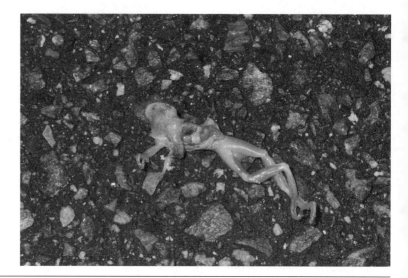

Roads can be deadly for individual frogs run over by cars and disrupt frog populations by fragmenting their natural habitat. Courtesy John D. Willson

nants, including heavy metals, and other chemicals found in road de-icing products, petroleum products, herbicides, and pesticides.

Because reproduction in most frogs and toads depends on the ability to communicate acoustically, anything that interferes with that communication can detrimentally impact frog and toad populations. A study conducted in Australia on two species of frogs found that one of the frogs, known as the southern brown treefrog (*Litoria ewingii*), increased the pitch of its call during periods of traffic noise, which is primarily a lower-pitched sound, thus increasing the ability of females to hear males in a noisy environment. Of course, a trade-off must then be made by males; that is, they must alter their call so that they can be heard by females but cannot afford to alter their call so much that they are no longer attractive to females.

Roads have an isolating effect on frog and toad populations that causes habitat fragmentation. Many amphibians rely on metapopulations to sustain their overall population levels. A metapopulation consists of several smaller populations among which individuals frequently move. If the ability to move among the subpopulations is interfered with, then the metapopulation dynamics break down and the overall population of frogs will decline and potentially become extinct. For many frog species, roads, especially highways, are impassable and often separate otherwise large, healthy metapopulations of frogs by preventing movement among the various subpopulations. Understanding how habitat fragmentation by roads affects frog populations is complex and requires considering a number of factors, including, but not limited to, habitat, type of road, and dispersal ability of the frog species. Extensive research is required to develop a more complete understanding of this issue, which will allow the fragmentation effects on amphibian populations by roads to be avoided or mitigated.

Frogs: The Animal Answer Guide

Naming roads after plants and animals is popular in many regions, but unfortunately, the presence of a road is seldom in the best interests of frogs and toads that have to cross them, no matter what the road is named. Courtesy Clyde Sorenson

Why are there no frogs living in the wetland or stream behind my house?

Many people remember having frogs and toads living in wetlands or streams behind their homes. One of us (MD) remembers finding toads breeding in pools in his backyard when he was a child only to find that, after 10 years, such toad populations no longer existed. A variety of possible reasons may explain the disappearance of frogs from people's backyards. First, over the years, as neighborhoods expand and shopping malls and other developments crop up, suitable, available frog and toad habitats decrease. In addition, chemicals from roads and other impervious surfaces may pollute the streams or wetlands, making reproduction difficult. Finally, increasing impermeable surfaces means that water within a stream's watershed that would normally percolate through the soil, eventually making it to streams and is dumped directly into the stream by way of storm sewers so that stream volume and velocity increase dramatically during heavy rains. Such increases in velocity can have a "flushing" effect on amphibian populations, displacing them and their eggs some distance downstream. One of us (MD) moved into his current house in a new neighborhood about 10 years before writing this book. At that time and for several years afterward, Cope's gray treefrogs (*Hyla chrysoscelis*), spring peepers (*Pseudacris crucifer*), and upland chorus frogs (*Pseudacris feriarum*) all called from the floodplain of the stream near his house. Gradually, the frequency and intensity of the choruses decreased until now, 10 years later, no frogs are heard. It is likely that reproduction, for the reasons listed previously, was greatly reduced in the stream's floodplain after the neighborhood was developed, and adult frogs were able to survive for a few years afterward with limited successful reproduction. After a few years, negative impacts from development reduced successful reproduction so that populations that once occurred there and called every night during spring and summer no longer exist. Watersnakes inhabit that stream as well and depend on frogs as a primary food source. Although watersnakes were never as commonly encountered as the frogs and toads were heard, it is likely that they too have declined or disappeared since the neighborhood has developed.

What can an ordinary citizen do to help frogs?

When we are faced with widespread environmental issues, such as declining and disappearing frog populations, the answers to fixing such problems are complex but depend ultimately on all of us. If we are concerned about frogs and their environments, then we can make enough of a difference to curtail or reverse the declines of frog populations. Aside from being environmentally conscious and friendly, perhaps we can inform ourselves about issues involved in stemming the decline of frogs and toads. Numerous organizations are available that will allow ordinary citizens to get involved in frog conservation. Such organizations as Partners in Amphibian and Reptile Conservation, the Audubon Society, and local herpetological societies have plenty for motivated citizens to do. The North American Amphibian Monitoring Program and FrogwatchUSA are great ways for citizen scientists to help monitor frog populations near their homes and to provide critical data for evaluating frog populations in their area and across the country.

Ordinary people can do much to help frogs. Make your backyard "frog friendly" by providing cover for frogs and toads and by avoiding using pesticides that kill insects (i.e., frog food) and frogs. If you are really motivated, you might even try to build a wetland in your backyard as a frog breeding habitat. The U.S. Department of Agriculture has online instructions for constructing backyard wetlands. Never introduce predatory fish into areas where frogs and toads breed, and never release nonnative frogs because they may carry disease.

Most important, become politically active as an educated citizen con-

PARC (Partners in Amphibian and Reptile Conservation) is a national and international organization whose members are dedicated to the conservation of frogs, toads, and other amphibians and reptiles.

cerned about frogs and toads. Politicians are often misinformed about the importance of wetland areas and about frogs and other animals. Concerned, motivated, persistent voters can have amazing impacts on the political process by getting legislators to pass laws and enforce regulations that help to protect frogs and their habitats. The Web site www.savethefrogs.com is a great resource for anyone interested in learning why frogs are disappearing and what we, as humans, can do about it. See Appendix B for a list of resources and organizations dedicated to frog biology and conservation.

Chapter 11

Frogs in Stories and Literature

What roles do frogs play in religion and mythology?

Placing frogs or toads in the context of religion is difficult because of evolving cultures and because the frog or toad has not assumed the lasting significance as the sole icon of a major civilization. Nonetheless, scholars have cited numerous religious roles for frogs. A few thousand years ago, Egyptians credited at least four gods with having frog-like heads and with being involved in the creation of the earth. Because the religions of Egypt have extended over more than 5,000 years, other animals have also assumed the deity roles, so frogs have not really been that special in Egyptian religion or mythology.

Frogs are actually mentioned in three books of the Bible (Old Testament: Exodus, Psalms; New Testament: Revelation). Frogs were not popular creatures of the times. The Lord's admonishment to Pharoah pretty well sums up the negative role frogs played in the Judeo-Christian religion: "And if thou refuse to let them go, behold, I will smite all thy borders with frogs" (Exodus 8:2, King James Version [KJV]). The last book of the Bible states, "And I saw three unclean spirits like frogs come out of the mouth of the dragon, and out of the mouth of the beast, and out of the mouth of the false prophet" (Revelation 16:13, KJV). Not a pretty sight.

One story about frogs that qualifies as an urban legend is about the boiling frog. A summary and analysis of the myth, written by one of us (JWG) for a newspaper column, follows:

The Legend of the Boiling Frog Is Just a Legend

Recently I received a communication about frogs that emphasizes the importance of confirming conventional wisdom and offers a metaphor for the human response to environmental degradation.

The issue started with an email from Germany. As often happens in scientific inquiry, though the answer to the question was pretty straightforward, arriving at the answer was not. But the easy way out—accepting what "everyone knows"—more often than not simply perpetuates misinformation. Although finding an answer that destroys an urban myth or a commonly held belief may disappoint some people, we are better off knowing the truth.

Joe Pechmann at the University of New Orleans, who is a noted amphibian conservation biologist, received a query last month that read:

"I am writing a weekly column for *Die Zeit*, Germany's major weekly paper, on scientific urban legends that my readers ask me about. Now you surely have heard the story of the boiling frog that is often told by consultants or activists: If you put a frog in boiling water, he will try to escape. If you put him in cold water and heat it gradually, the frog will remain in place until he's boiled, because that's the lesson, to him (and consequently to us)—gradual change is not perceivable. Frankly, I don't buy this. But I am looking for professional advice (and I don't want to boil frogs). Can you help me with that question? Thanks! Christoph Droesser, Hamburg, Germany."

Joe was not sure what the answer was, so he referred Mr. Droesser to me. I also passed the buck, saying: "I have heard the anecdote many times and actually heard a Baptist preacher give a sermon in Mississippi in which he used the story of a big bullfrog in a bucket of water that was being heated. The situation was presented as an example of how gradual habituation to a devilish situation leads to acceptance of an even worse one. But with a real frog in real water, my bet is that when it began to get uncomfortable the frog would jump out if it could, long before the water started to boil. Nonetheless, consultants, activists, and others who are unaware of gradual environmental problems are responding in the way we like to think a frog acts rather than the way it does."

I went on to say, "Although I do not know a databased answer myself, I am aware of experiments that have been done on responses of amphibians to thermal conditions. In some of the experiments the temperature was gradually raised, so I feel certain someone familiar with those studies would have an impression of what a frog would do as the water warmed up. I am sending your question to Dr. Victor Hutchison at the University of Oklahoma to see what he says. I would be interested to know also."

Vic's answer was as follows: "The legend is entirely incorrect! The 'critical thermal maxima' of many species of frogs have been determined by several investigators. In this procedure, the water in which a frog is submerged is heated gradually at about 2 degrees Fahrenheit per minute. As the temperature of the water is gradually increased, the frog will eventually become more and more active in attempts to escape the heated water. If the container size and opening allow the frog to jump out, it will do so." Naturally, if the frog were not allowed to escape it would eventually begin to show signs of heat stress, muscular spasms, heat rigor, and death.

So where does that leave us with the metaphor for the human response to environmental degradation? Well the idea that you can induce a frog to remain in boiling water if you start it off in cold water is not true biologically. But that does not diminish the need to keep an eye out for the gradual relaxation of environmental laws and regulations. The metaphor lies in the frog's ability to escape from the container: if there's no way out, then the frog's fate is a foregone conclusion.

What do frogs have to do with fiction or fairy tales?

Among the fairy tale classics that deal with frogs is "The Frog King or Iron Heinrich" credited to the Brothers Grimm (Wilhelm and Jacob) of Germany in the 1800s, although similar folktales can be found in other cultures. Even the Grimms's story has been retitled "The Frog Prince" in modern versions. Numerous translations have led to variations of what transpires between the princess and the frog she finds in a pond, but the basic premise is the same. The frog is actually a prince on whom a witch (a wicked one, of course) has cast a spell that can only be broken if he is befriended by a princess and either kissed, fed, allowed to sleep with her, or all three. In some versions, the princess kisses the frog, who immediately appears as a prince in full regalia. In the original version, the princess was not portrayed in such a compassionate light. Instead, near the end of the story, she is so annoyed with the frog's insistence on being a part of her life that she hurls him against the wall, whereupon he falls to the floor and turns into a prince. The moral is the same for any of these stories: within some frogs is a prince, so be careful not to judge too critically. Iron Heinrich, by the way, was the prince's former coachman who could not care less about the princess but was thrilled that his master was no longer a frog.

Arnold Lobel's children's book *Frog and Toad Are Friends* is a simply written and illustrated but wonderful story of Frog and Toad that begins as winter ends and spring begins. The different but likable personalities of the two friends and the simple plots of the stories make this enjoyable reading as a bedtime story for children. The following excerpt is an example of the

Frogs: The Animal Answer Guide

outlook on life of the lively frog and lethargic toad, presentations that fit what we might think their personifications should be.

Spring
Frog ran up the path
to Toad's house.
He knocked on the front door.
There was no answer.
"Toad, Toad," shouted Frog,
"wake up. It is spring!"

"Blah," said a voice
from inside the house.

Frogs made it into Aesop's Fables in stories about frogs looking for a leader, frogs making rabbits think things could be worse, and a frog that became too full of himself. None of them are particularly complimentary about frogs. Aesop apparently did not recognize frogs as distinctive from toads (or a translator somewhere between the Greek and the current versions of the stories did not), based on the three fables that follow.

The Frogs Desiring a King
The Frogs were living as happy as could be in a marshy swamp that just suited them; they went splashing about caring for nobody and nobody troubling with them. But some of them thought that this was not right, that they should have a king and a proper constitution, so they determined to send up a petition to Jove to give them what they wanted.

"Mighty Jove," they cried, "send unto us a king that will rule over us and keep us in order."

Jove laughed at their croaking, and threw down into the swamp a huge Log, which came down splashing into the swamp. The Frogs were frightened out of their lives by the commotion made in their midst, and all rushed to the bank to look at the horrible monster; but after a time, seeing that it did not move, one or two of the boldest of them ventured out towards the Log, and even dared to touch it; still it did not move. Then the greatest hero of the Frogs jumped upon the Log and commenced dancing up and down upon it, thereupon all the Frogs came and did the same; and for some time the Frogs went about their business every day without taking the slightest notice of their new King Log lying in their midst.

But this did not suit them, so they sent another petition to Jove, and said to him, "We want a real king; one that will really rule over us." Now this made Jove angry, so he sent among them a big Stork that soon set to work gobbling them all up. Then the Frogs repented when too late.

Moral of story: Better no rule than cruel rule.

The Hares and the Frogs

The Hares were so persecuted by the other beasts, they did not know where to go. As soon as they saw a single animal approach them, off they used to run. One day they saw a troop of wild Horses stampeding about, and in quite a panic all the Hares scuttled off to a lake hard by, determined to drown themselves rather than live in such a continual state of fear. But just as they got near the bank of the lake, a troop of Frogs, frightened in their turn by the approach of the Hares scuttled off, and jumped into the water. "Truly," said one of the Hares, "things are not so bad as they seem."

Moral of story: There is always someone worse off than yourself.

The Frog and the Ox

"Oh Father," said a little Frog to the big one sitting by the side of a pool, "I have seen such a terrible monster! It was as big as a mountain, with horns on its head, and a long tail, and it had hoofs divided in two."

"Tush, child, tush," said the old Frog, "that was only Farmer White's Ox. It isn't so big either; he may be a little bit taller than I, but I could easily make myself quite as broad; just you see." So he blew himself out, and blew himself out, and blew himself out. "Was he as big as that?" asked he.

"Oh, much bigger than that," said the young Frog.

Again the old one blew himself out, and asked the young one if the Ox was as big as that. "Bigger, father, bigger," was the reply.

So the Frog took a deep breath, and blew and blew and blew, and swelled and swelled and swelled. And then he said: "I'm sure the Ox is not as big as this." But at this moment he burst.

Moral of story: Self-conceit may lead to self-destruction.

Among the fairy tales Hans Christian Andersen told was "The Toad," written in 1866. The story begins in a well where both frogs and toads live, the toads being by far the ugliest of beings and not particularly welcomed by the frogs, as seen by the comment about the mother-Toad.

"She's thick, and fat and ugly," said the young green Frogs; "and her children will be just as ugly as she is."

"That may be," retorted the mother-Toad, "but one of them has a jewel in his head, or else I have the jewel."

The young frogs listened and stared; and as these words did not please them, they made grimaces and dived down under the water. But the little Toads kicked up their hind legs from mere pride, for each of them thought that he must have the jewel; and then they sat and held their heads quite still.

Frogs: The Animal Answer Guide

Meanwhile, the ugliest of the young toads has a wanderlust to see the world above the well, and she manages to escape in the bucket as it is drawn to the top by a farm laborer. After a few adventures in the area, the young toad eventually overhears a stork family talking of adventures even farther away.

And the Mother-Stork began talking in the nest, and told about Egypt and the waters of the Nile, and the incomparable mud that was to be found in that strange land; and all this sounded new and very charming to the little Toad.

"I must go to Egypt!" said she. "If the Stork or one of his young ones would only take me! I would oblige him in return. Yes, I shall get to Egypt, for I feel so happy! All the longing and all the pleasure that I feel is much better than having a jewel in one's head."

And it was just she who had the jewel. That jewel was the continual striving and desire to go upward—ever upward. It gleamed in her head, gleamed in joy, beamed brightly in her longing.

We, of course, will not spoil the ending for future readers, but the lesson is that any of us inspired and motivated to learn, travel, and appreciate the world, are already possessors of the toad's jewel.

What roles do frogs play in native cultures?

Frogs and toads have been the focus of legends, myths, and religion over the centuries in many different native cultures of the world. Some stories have probably been embellished in modern writings and their significance to a people inflated, depending on the writer and the culture involved. But clearly these animals were a part of the environment and daily life in many areas. Frogs and toads indeed appear in the artwork of tapestries, pottery, and wall paintings from numerous countries throughout the world.

An excellent overview of the role of frogs or toads in cultures as diverse as the Aztecs, Egypt, Greece, and China, is given by Ellin Beltz in *Frogs: Inside Their Remarkable World.* The belief that frogs symbolized fertility extends from ancient Egypt, the Greek island of Crete 4,000 years ago, to Native Americans of the West and other cultures. Some species lay thousands of eggs and sometimes thousands of recently metamorphosed froglets or toadlets can cover the ground in an area. An ironic connection to twentieth-century medicine is that certain species of frogs or toads were commonly used in pregnancy tests.

Beltz also tells a delightful story of a legend among the Kootenai tribe in northern Idaho. Antelope challenges Frog to a race, with both animals betting heavily on their success. Of course, an antelope can outrun any frog, but the chief of the frogs has a plan. He ordered all of the frogs in the village to space themselves a few feet apart from the starting line to the finish line. Then, as Antelope approached, the next frog in front of him would hop. Antelopes, like most people, cannot tell one frog within a species from another, so the perception was that the Frog was outhopping Antelope, and Frog wins the race.

Based on her research into Asian culture, Beltz tells a tale in her book about a Chinese legend in which "a woman named Chang O stole the Elixir of Immortality from her husband and fled to the moon." Other things happen to Chang O; she "was changed into a three-legged frog and forced to stay on the moon." Our interpretation is that the Elixir of Immortality would not be so desirable if you end up as an immortal frog with only three legs.

What roles do frogs play in modern or popular culture?

Frogs and toads have had their share of appearances in commercials, songs, and festivals in modern times. A national commercial in the United States, first shown to hundreds of thousands of viewers during Super Bowl XXIX in 1995, traded on the phenomenon that male frogs call. Three large, deep-voiced bullfrogs in the commercial repeat their chorus as "Bud," "Weis," and "Er," as they sit atop rocks in a stream in front of a tavern with a Budweiser sign.

Kermit the Frog is perhaps the most famous of Jim Henson's Muppets and appeared first in 1955. Already a star of television through *Sesame Street*, Kermit became even more popular for his behavior and songs in *The Muppet Movie* in 1979. Another popular frog cartoon character who also first appeared in 1955, in Looney Tunes *One Froggy Evening*, was Michigan J. Frog. Wearing a top hat and carrying a cane, both of which he handles like a vaudeville stage star, the ever-so-happy frog sings ragtime and other songs.

Frogs and toads have also found their way into the music world as both the topic of songs and the names of bands. One of the best-known examples is the star in the song "Joy to the World," Jeremiah, a bullfrog. The song was written by Hoyt Axton, released by Three Dog Night in 1979, and made its way to the number one spot in popular music in 1979. The opening line, "Jeremiah was a bullfrog," is known by most followers of popular music. A band that had two songs reach the top 20 based on *Billboard* magazine's national ratings was known as Toad the Wet Sprocket.

Frogs are not as popular as flowers or food for community festivals, but at least three places in the world have frog or toad races. The Calaveras

County Fair & Jumping Frog Jubilee was popularized by Mark Twain's story. The jubilee is held each year during the third week in May and features a frog-jumping contest (the record is more than 20 feet). Australia has taken the Calaveras County frog-jumping races to a new level, a lower one actually, with the advent of cane toad racing in many areas, nearly always associated with bars. A typical version of the contest is simple, with simple prizes (beer), in which invasive cane toads are collected from the local area and given various names that have some relationship to the people in the crowd. Toads are covered by a bucket on a dance floor (no dancing allowed during the race), and then set free to race (that is, hop, often very slowly) toward the edge of the dance area amid hoots, howls, and cheers from an inspired crowd. This is about the only popular activity that cane toads in Australia engage in because of their destructive nature both as predator and poisonous prey. The recommendation by some bars is that both the winners and losers (the toads, not their sponsors) be euthanized (usually by being placed in a bag in a deep freeze) and that new contestants be caught for the next night's races.

Conway, Arkansas, promises everyone will have a "toadally good time" at the annual, free-admission Toad Suck Daze festival that features the Toad Race competition. Children are encouraged to capture their own toads in the surrounding area and enter them in the races. The toad that responds best to loud contestants and spectators shouting and yelling presumably wins the race. After the race, toads are released at the sites where they were captured. The community uses the honor system, and testing for steroids, human growth hormones, and other athletic enhancement drugs has not been implemented.

Frogs, a so-called horror movie, was released in 1972 that featured an attack on an ecologically insensitive Florida mansion owner by frogs, as well as tarantulas, snakes, and birds. The film did not win any Academy Awards but did receive a Tomatometer rating of 21 percent on the Rotten Tomatoes film critics site. The ratings could have been worse, but the movie poster showed a large bullfrog-like creature with a human arm hanging out of its mouth, which may have kept its Tomatometer rating out of the teens.

The documentary *Cane Toads: An Unnatural History* was released in 1988 and documents the spread of the invasive cane toad in Australia. The movie is accurate but is presented in such a humorous and quirky way that it is a joy to watch. One tagline says, "If Monty Python produced a National Geographic Special, it would be *Cane Toads*!"

Many jokes have been told in which frogs or toads are primary characters or unwitting victims. Over the years, we have heard and told many such jokes. The joke that follows provides some insight into our humor.

A poster from the 1972 movie *Frogs* implied that man-eating frogs would soon be invading the world. Courtesy Margaret Wead

We will let the reader decide whether it is actually funny.

> A frog walks into a bank to get a loan. He approaches the counter and the teller says, "My name is Patty Black. How can I help you?" The toad replies, "I need a loan." The teller then asks the frog what he intends to use as collateral. The toad pulls out of his pocket a small ivory statue and hands it to the teller. The teller says, "I'll have to ask my manager about this" at which point she turns and walks into the bank manager's office. She explains the situation and shows her manager the little ivory statue. After careful thought, the bank manager says, "That's a knickknack, Patty Black, give the frog a loan."

What roles have frogs played in poetry and other literature?

Frogs and toads have had their place in poetry and prose in many languages and cultures. One of the most delightful fictional amphibians ever concocted was in Kenneth Grahame's book *The Wind in the Willows*

Frogs: The Animal Answer Guide

published in 1908. This book is full of humanized animal characters, but the protagonist in the story is Mr. Toad of Toad Hall. Mr. Toad is a loud, too-full-of-himself, overbearing, tempestuous creature who squanders his wealth, drives too fast and recklessly, and leads a turbulent life. But he is a good-hearted toad whose closest friends are Mole, Ratty, and Mr. Badger, all of whom spend time getting Mr. Toad out of trouble. The essence of Mr. Toad and his own lofty attitude about himself is evident in the last part (chapter 12) in which he likens his appearance at Toad Hall to the "The Return of Ulysses" with a song:

> The Toad-came-home!
> There was panic in the parlours and howling in the halls,
> There was crying in the cow-sheds and shrieking in the stalls,
> When the Toad-came-home!
> When the Toad-came-home!
> There was smashing in of window and crashing in of door,
> There was chivvying of weasels that fainted on the floor,
> When the Toad-came home!
> Bang! go the drums!
> The trumpeters are tooting and the soldiers are saluting,
> And the cannon they are shooting and the motor-cars are hooting,
> As the—Hero—comes!
> Shout—Hoo-ray!
> And let each one of the crowd try and shout it very loud,
> In honour of an animal of whom you're justly proud,
> For it's Toad's—great—day!"

The Wind in the Willows was presented in 1929 as a play, *Toad of Toad Hall*, written by A. A. Milne, the author of *Winnie the Pooh*. In 1949, the story became further popularized as an animated movie by Walt Disney in *The Adventures of Ichabod and Mr. Toad*. Mr. Toad's adventures were distinct from Ichabod Crane's, which were based on "The Legend of Sleepy Hollow." Mr. Toad's Wild Ride subsequently became one of the original rides both at the opening of Disneyland in Anaheim, California, and Disneyworld's Magic Kingdon in Orlando, Florida, in 1971.

The single-most famous frog in fictional prose is probably one whose name, Daniel Webster, is not as well known as the author, Mark Twain, or the name of the tale itself, *The Celebrated Jumping Frog of Calaveras County*. Although we will not spoil the ending of Mark Twain's story for those who have not read it or cannot remember what happened when the frog's owner and another man place a wager on seeing which frog could outjump the other, an impression of Daniel Webster's prowess and temperament can be gained from the following passage:

Why, I've seen him set Dan'l Webster down here on this floor—Dan'l Webster was the name of the frog—and sing out, "Flies, Dan'l, flies!" and quicker'n you could wink he'd spring straight up and snake a fly off'n the counter there, and flop down on the floor ag'in as solid as a gob of mud, and fall to scratching the side of his head with his hind foot as indifferent as if he hadn't no idea he'd been doin' any more'n any frog might do. You never see a frog so modest and straightfor'ard as he was, for all he was so gifted.

From *The Saturday Press*, Nov. 18, 1865. Republished in *The Celebrated Jumping Frog of Calaveras County, and Other Sketches (1867)*, by Mark Twain, all of whose works are published by Harper & Brothers.

This tale took place at a gold-mining camp in Angels in southwest Calaveras County, which is southeast of Sacramento. On the basis of its location and that the frogs were caught in the vicinity in a muddy, swampy habitat, we assume that the frogs in the story were presumably California red-legged frogs (*Rana aurora draytoni*) as the bullfrog had not yet become established as an invasive species in the region. Ironically, red-legged frogs cannot be used in the jumping frog contests of today because the species is on the federal endangered species list.

This coin celebrates the Calaveras County Fair & Jumping Frog Jubilee, made famous by Mark Twain.

Courtesy C. Kenneth Dodd

Frogs: The Animal Answer Guide

Perhaps the most famous poet to write about a frog, and not in a particularly flattering way from the standpoint of an introvert, was Emily Dickinson. One version of what she wrote is,

I'm nobody, who are you?
Are you nobody too?
There's a pair of us, don't tell!
They'd banish us, you know!

How dreary to be somebody!
How public like a frog,
To tell your name the livelong day
To an admiring bog!

A famous American writer, E. B. White, who wrote about a mouse (*Stuart Little*) and a spider (*Charlotte's Web*), did not write a book about frogs and toads but did state that "analyzing humor is like dissecting a frog. Few people are interested and the frog dies of it." Being a humorist, he might have even said, "Few people are interested and the frog croaks."

Chapter 12

"Frogology"

Who studies frogs?

Many laypeople are interested in and enjoy studying and learning about the biology of frogs and other animals to increase their appreciation and understanding of the natural world. Scientists usually study frogs either to learn about frogs or to learn about a general scientific phenomenon, using frogs as model organisms. Most scientists who study frog biology refer to themselves as herpetologists. In general, herpetologists conduct scientific research on amphibians and reptiles (i.e., snakes, turtles, lizards, crocodilians, salamanders, frogs), but many herpetologists focus their studies on only one group and, in some cases, on only one species.

Recognizing declines in frog and toad populations worldwide since the 1980s has dramatically increased the numbers of scientists studying these species over the past few decades. Numerous scientists have begun investigating various causes of frog and toad declines. Mike Lannoo of Indiana State University has been at the forefront of much of the research conducted in North America on amphibian declines and has authored or edited several books, including *Amphibian Declines*, a massive tome that details the conservation status of every species of amphibian in the United States. Lannoo also authored *Malformed Frogs: The Collapse of Aquatic Ecosystems*, which focuses on the potential importance of frogs with gross malformations and what these malformations can tell us about the ecological integrity of our environment. Ray Semlitsch of the University of Missouri has studied amphibians during his entire career and recently edited *Amphibian Conservation* in which experts in various fields address critical issues related to the ecology and conservation of amphibians. Karen Lips at the University of Maryland has focused much of her work on understanding factors that can cause declines in amphibian populations, especially chytrid fungus that has annihilated many frog populations in the American Tropics. Lips and her collaborators have been at the forefront of showing how severe declines in frog populations from disease can have cascading effects on other parts of tropical ecosystems.

Some herpetologists focus on understanding the evolution and diversity of frogs and toads. Typically, these people consider themselves systematists or taxonomists, and they focus on understanding the variations found among many types of frogs and toads and how they should be recognized and grouped as species so that the groupings match their evolutionary history. Most modern systematists use genetic characteristics as their primary tool for understanding relationships among species, although earlier morphological studies used to differentiate among species, genera, and families have been extremely useful. Many new species, primarily from the Tropics worldwide, are still described each year. Unfortunately, in many cases, new species are already on the verge of extinction. Many frog and toad species have already gone extinct because of human-induced causes scientists are now discovering.

Numerous resources are available to the layperson and scientist to aid in studying the biology of frogs. Darrel Frost of the American Museum of Natural History has been a leader in the field of evolutionary relationships among frog groups and has taken the lead on a project called Amphibian Species of the World, an online project (http://research.amnh.org/vz/her petology/amphibia/) that strives to maintain up-to-date information about diversity and evolutionary patterns of amphibians so that professionals can use information for conservation and other research purposes. David Wake at the University of California at Berkeley coordinates another major online project called AmphibiaWeb (www.amphibiaweb.org), which provides detailed information on the biology and conservation of many known species of amphibians worldwide.

Why do scientists study frogs?

People study frogs for a variety of reasons. Some scientists study frogs because they are interested in frog biology and want to understand how they live and respond to their environment. Such information is critical when developing effective conservation plans for frogs, and, unfortunately, for many species, we know so little about their basic biology that developing methods to ensure their conservation is difficult. Most scientists that study frogs only study one aspect of their biology. For example, ecologists that study frogs focus most likely on field studies of frogs in natural environments. Physiologists that study frogs specialize on examining how the frog's body works and how it responds to external conditions. Many scientists may not study frogs because they are primarily interested in frogs, but they are instead exploring general phenomena for which frogs make good study organisms. As explained in detail elsewhere in this book, frogs can serve as model organisms for the study of many different general biological phenomena.

Some people are interested in learning about frogs so that they can keep them in captivity. Herpetoculturists, or people that keep and breed frogs (and other reptiles and amphibians) primarily for pets, learn as much as they can about their biology and reproductive characteristics to maximize the chances the animals will thrive and breed in a captive environment such as a terrarium. Such research is important, and because declines in frog populations in the wild have become widely recognized as a major problem, many zoos and other organizations are working together to keep and breed species of frogs that may now be extinct in the wild. The hope is that, eventually, these species can be returned to the wild. Dante Fenolio and his colleagues at the Atlanta Botanical Garden, in partnership with Joe Mendelson at Zoo Atlanta, have developed a program that uses the expertise and facilities at the botanical garden and at Zoo Atlanta to promote education and conservation of imperiled frogs.

How do scientists study frogs?

Scientists use various techniques to study frogs and toads. These techniques have been developed over the years and include some fairly basic and, in some cases, sophisticated approaches. Frogs are used in laboratory experiments, partly because some species can be easily maintained in captivity, thus making excellent subjects for laboratory studies of their physiology and behavior. A significant aspect of conducting laboratory experiments on any animal is that specimens are able to be acquired efficiently and with minimal cost. Tadpoles are usually easily captured with a dip net and are able to be maintained in aquaria or aquatic holding tanks. Many species of frogs and toads can be successfully bred in captivity as well as easily captured in the field. Vocal animals like frogs and toads are also ideal animals for studies that focus on auditory communication in animals.

An effective way to capture large numbers of frogs and toads for scientific research is to use a drift fence. The fence is simply a "wall" of aluminum sheeting or silt fencing erected where frogs and toads occur (such as near a wetland). Funnel traps or buckets sunk in the ground are strategically placed at intervals alongside the fence. When amphibians encounter the drift fence and cannot get over it, they will crawl or hop along the

The frog facility at the Atlanta Botanical Garden in Atlanta, Georgia, houses several species of rare frogs, some of which may be extinct in the wild. The research programs investigate aspects of the life history, ecology, and reproduction of some species to determine whether they can be raised successfully in captivity. Courtesy Dante Fenolio

fence, eventually finding themselves captured in the funnel trap or bucket. Drift fences are frequently placed between a wetland breeding site and the upland habitat where the species spends most of the year so that animals coming in to breed are captured. A drift fence study at Rainbow Bay on the Savannah River Site in South Carolina has been ongoing since 1978. The drift fence and traps have been checked every day since the fence was installed, making it the longest running study of amphibians anywhere in the world. More than 60 scientific articles have been published as a result of this study. This long-term study has been critical in evaluating factors related to amphibian declines and will provide a baseline for studies of climate change.

Because many frogs and toads call at somewhat predictable times, several research approaches take advantage of the distinctive calls of different species to locate individuals or choruses at night. Organized calling surveys in which volunteers record times, locations, and weather conditions when certain species are breeding has provided valuable information on the patterns of persistence and declines of rare and common species of frogs. The

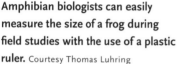

Amphibian biologists can easily measure the size of a frog during field studies with the use of a plastic ruler. Courtesy Thomas Luhring

North American Amphibian Monitoring Program (NAAMP), a national program, is operated by Linda Weir of the U.S. Geological Survey (USGS) at the Patuxent Wildlife Research Center in Maryland. NAAMP is a collaborative effort created in 1997 between USGS and regional partners, including state natural resource agencies and nonprofit organizations. The stated goal is "to monitor the status and trends of calling anuran populations of eastern and central North America through standardized surveys." The approach is for people to listen for frog and toad calls at wetlands across the country to determine whether species are declining, increasing, or remaining stable. Resulting data could influence environmental management decisions, and the concept has potentially positive features from a conservation perspective. If done properly, with the rigor of a scientific study, such surveys can measure when and where frogs are thriving. Doing this year after year will help to indicate that certain species are less common in certain areas than was previously thought.

"Frogloggers" are automated recording systems that can be left for weeks to record frog calls without maintenance at remote locations. This system allows scientists to study factors affecting the calling patterns of frogs by automatically recording frog calls at regular intervals. Researchers then download the froglogger at periodic intervals and listen to the calls to study the behavior of the frogs. A study by Charlotte Steelman at Davidson College showed that frogs actually reduce calling during noisy periods when airplanes flew over a wetland. A study by Andrew Bridges in

Frogs: The Animal Answer Guide

South Carolina revealed that leopard frogs, which were never heard calling during early evening hours at his study site and were thought to be absent, were actually present but called only after midnight, as revealed by frogloggers.

Eggs of some species (or at least groups) of frogs are fairly distinctive, and surveying wetlands in a systematic way for egg masses or strings of eggs is an effective way to evaluate breeding in some frog species. Because many species of frogs lay thousands of eggs at a time, most of which never survive, researchers can remove eggs for use in experiments without causing problems for the local frog population. Similarly, studies of tadpoles, which are easily captured using dip nets or minnow traps, can reveal much about the ecology of frog and toad populations.

A standard way of studying frogs, toads, and many other animals is to conduct a capture-mark-recapture study. Frogs are captured, measured, marked, and then released at the site of capture. Later, if recaptured, the researchers can determine growth rates and distances moved between captures. Mark-recapture data can also be used to estimate population sizes, survival, and recruitment rates. Frogs and toads can be marked for individual identification or cohort marked (i.e., all marked in the same way). Historically, the most widespread marking technique is to clip various combinations of toes so that when recaptured, individuals can be distinguished based on which toes are shorter than the others. This technique must be used with caution on some species so that toes critical to the frog's survival are not removed. For example, removing too many toes from a treefrog would likely diminish its ability to climb and threaten its survival. Another less-invasive marking technique is using fluorescent-colored elastomers, which are injected as a liquid under the frog's transparent skin. The elastomer hardens to the consistency of silicone and forms a marker visible under fluorescent light. Different combinations of colors, locations, and configurations can be easily used to provide the researcher with hundreds of unique marks that can be used individually to distinguish many specimens. Shannon Pittman, a student at Davidson College, marked Cope's gray treefrogs by injecting tiny, fluorescent tags with numbers on them under the skin of the leg of the frog. The tags could be easily read through the frog's translucent skin using an ultraviolet light and magnifying glass. Many amphibian biologists have also used passive integrated transponders, or PIT tags, to mark frogs and toads. PIT tags are microchips injected into the body cavity with a large needle and then later read by a special reader. The reader creates an electromagnetic field that charges the coil in each PIT tag causing it to transmit a signal (i.e., ID number) back to the reader allowing the scientist to identify the frog individually. Such devices have been used for years to identify pet dogs and cats.

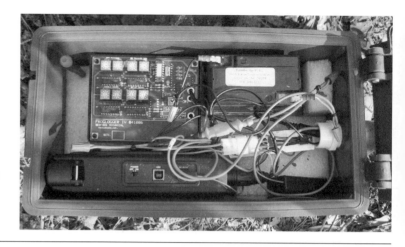

Automated recording systems, "frogloggers," can be left at remote locations to determine the calling patterns of frogs. Courtesy Charlotte Steelman

The best method for studying the movements and activities of wild frogs is to use radiotelemetry. Although radiotransmitters are often too large for small species of frogs, radiotelemetry has been used to effectively study the behaviors of several species of free-ranging frogs and toads in the field. Typically, scientists attach a small radiotransmitter to the animal's back, and it emits a signal at a particular radio frequency that the researcher can follow with a receiver and directional antenna. Radiotelemetry is effective for studying frog activity, movement patterns, and habitat use. To track small frogs that cannot carry a transmitter or large ones for just a short time, some scientists dust their research animals with fluorescent powder. After release, the powder falls off the frog as it hops around making a trail that can be tracked using an ultraviolet light.

Sophisticated computer software is now frequently used to study frogs and toads. Geographical information systems (GIS) allow detailed analyses of frog populations and multiple spatial scales (local to continental) and have revolutionized how many animal studies are conducted in the field. Steve Price, a researcher at Davidson College, has conducted numerous studies examining the distributions of frogs using GIS and found that habitat alteration by humans can affect populations of frogs at local and at broad, regional scales. Such information is critical when developing plans for conservation and preservation of habitat.

In 1994, a major handbook of techniques was edited by Ron Heyer and his colleagues and published by the Smithsonian Institution Press. This handbook provided important information about techniques for amphibian study and has helped to shape many conservation-oriented investigations of frogs and toads. A more recent handbook, edited by Ken Dodd and published in 2010 by Oxford University Press, has chapters written by numerous experts in the field that provide critical information on directions and approaches for future amphibian research.

Frogs: The Animal Answer Guide

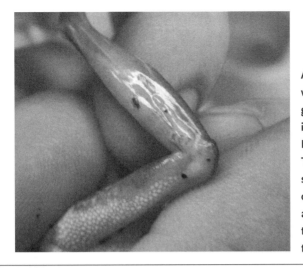

A numbered, flexible plastic tag is visible through the skin of a Cope's gray treefrog (*Hyla chrysoscelis*) being studied in a research project at Davidson College, North Carolina. The numbering system allows researchers to determine the identity of individual frogs so that when they are recaptured their movement patterns in the wild can be tracked over time. Courtesy Shannon Pittman

Which species are best known?

The African clawed frog (*Xenopus laevis*) has probably been studied more than any other frog species. For decades, scientists have used this species for laboratory experiments, and many articles have been written about its physiology, development, and biology. However, the number of studies conducted on this species in the wild are limited, and although it has been the focus of more scientific research than probably any other species of frog, we still know relatively little about many aspects of its ecology in its native habitat.

Typically, common species easily manipulated and maintained in captive situations are most well known. Numerous studies have been conducted on common species of toads (genus *Bufo*) and leopard frogs (*Rana pipiens*) to understand their ecology and behavior. The common European toad (*Bufo bufo*) has been studied extensively in its native range. Treefrogs (family Hylidae) are frequently studied and thus much is known about many of the common species.

Which species are least known?

Many species of frogs are so poorly known that only their basic biology is understood. Many species with limited geographic ranges or those that occur in difficult-to-reach regions have had no research conducted on them except for the basic description of the species. Many new species of frogs are discovered frequently, especially in the Tropics, and little is known about the extent of their geographic range. One could certainly argue that all the species of frogs yet to be discovered are the least known of all frogs and toads. That is, some are so poorly known, we do not know of their existence. Some widespread species remain relatively understudied because of their secretive habits. The eastern spadefoot toads (*Scaphiopus holbrookii*) occur throughout the southeastern United States. However, some aspects of its ecology are difficult to study because it may only emerge from underground retreats for a few days a year. Given the myriad unique and fascinating features that known frogs exhibit, it will be interesting to see what new and exciting aspects of frog biology have yet to be discovered.

How do scientists tell frogs apart?

Different species of frogs are distinguished from one another based on a number of characteristics, including size, shape, color, and pattern. One of the most critical determining factors when trying to identify a frog is to know where it is from. Because at any given location in the world, the number of species of frogs is limited, knowing which species are likely to occur in any particular area greatly narrows down the possibilities. For example, in North Carolina, there are 30 species of frogs and toads, some of which look very similar. But, if someone has a toad that is approximately 3 inches long, found in the Coastal Plain of North Carolina, then it is likely either a southern toad (*Bufo terrestris*) or a Fowler's toad (*Bufo fowleri*). To distinguish between these two somewhat similar species, a scientist would simply look for protruding knobs on the ridges (called cranial crests) between the toads eyes. Southern toads have such protrusions. In the Tropics, the numbers of species of frogs and toads increase substantially, and telling apart species with similar characteristics becomes more difficult.

Most species of frogs can be distinguished based on characteristics of their advertisement call, and most amphibian monitoring involves listening for frog and toad calls. Because each species in an area has a distinct call, a species can be identified even if it is not seen or captured. The calls of closely related species often, but not always, sound similar to one another, but there are frequently subtle differences between them that allow a trained ear to distinguish individual species. Lang Elliot, Carl Gerhardt, and Carlos Davidson's *The Frogs and Toads of North America* (2009) includes recorded calls on a CD, allowing the listener to practice identifying calls even while driving.

The characteristics of major groups of frogs (e.g., families) help set them apart from all other frogs. Many characteristics are skeletal and are thus difficult to ascertain in a living animal, but most groups have overall appearances that allow most trained scientists at least to assign them to the proper family. Some groups, for example, all frogs in the family Hylidae, have a small cartilage between the most distal digit of their toes that is generally easy to see.

Although tadpoles of some frog species characteristically have features

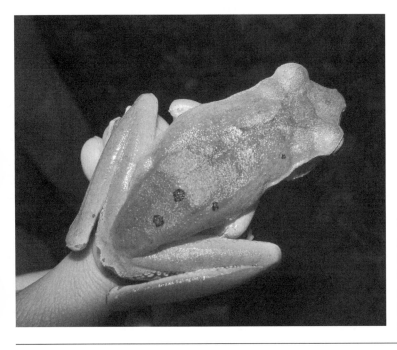

The yellow treefrog (*Polypedates pseudocruciger*), first described to science in 1998, is a large treefrog found only in the Western Ghats of India. Note the adult human's hand holding this large treefrog.

Courtesy Cris Hagen

that allow for relatively easy recognition, tadpoles of many species can be difficult to tell apart, even for many trained herpetologists. Young tadpoles are even more difficult to distinguish because they generally have not developed any characters that allow separation among possible species. One way to identify tadpoles is to examine their mouthparts, which are often distinctive for each species. Features such as the number of labial tooth rows or whether an oral disk is present are subtle characteristics but allow for positive identification. However, for some species, the tadpole is quite remarkable. In the southeastern United States, a solid black, 6-inch-long tadpole with bright red eyes should be as easy to identify as the tadpole of the river frog (*Rana heckscheri*). *Tadpoles: The Biology of Anuran Larvae*, published in 1999 by the University of Chicago Press and edited by Roy W. McDiarmid and Ronald Altig, has more than 450 pages that help to identify and understand the differences among this intriguing life stage of many frogs and toads.

As described elsewhere in this book, many frog biologists distinguish among species not by using morphological characters but, instead, by examining the genetic code of individual frogs and using that information to assign them to species. Of course, if you are wading around in a wetland, trying to identify the frog you just captured before you release it, conducting in-depth studies of its DNA is not practical.

The dorsolateral ridge that begins behind the eye and extends two-thirds of the way down the body distinguishes the green frog (*Rana clamitans*) from other closely related species, such as the bullfrog (no dorsolateral ridge) and leopard frog (complete dorsolateral ridge).

Courtesy John D. Willson

Why do the scientific names and classification of frogs change frequently?

Taxonomy is the science of naming organisms, and it is not a perfect science because new species are discovered and earlier misinterpretations of relationships are uncovered among species and genera. Taxonomists that work with frogs and toads try to classify and name them to reflect evolutionary relationships among species and groups of species. Systematists study the evolutionary relationships among groups of animals. Because taxonomists try to name frogs to match their evolutionary relationships, the fields of taxonomy and systematics are often difficult to distinguish.

If species are placed together within a genus (plural = *genera*), they are considered closely related, and closely related genera are grouped together to form a family. All animals are given a two-part scientific name. The first word of this scientific name is the name of the genus (e.g., the genus for humans is *Homo*), and genera are always capitalized. The second part of a scientific name is its species name (for humans that name is *sapiens*). Together, the genus and species make up the full scientific name (*Homo sapiens*). The species name of the upland chorus frog is *Pseudacris feriarum* and that of the southern chorus frog is *Pseudacris nigrita*. Putting both species in the same genus indicates that they are more closely related to each other than they are to frogs that might be in the same family but are in other genera, such as *Hyla* or *Smilisca*.

Taxonomists frequently disagree about the importance of different traits in classifying animals, and scientists that work on frog taxonomy are no exception. Modern molecular genetics has helped many herpetologists

Frogs: The Animal Answer Guide

to resolve taxonomic questions but have also created additional questions, some of which can be contentiously argued. Oftentimes, taxonomists will recommend changing species' scientific names to reflect the most accurate understanding of the natural groupings of frogs. Because of disagreements about the evolutionary relationships among groups of species and because neither of us is a taxonomist, we have not provided a definitive list of the major groups (e.g., families) of frogs and toads. Instead, we refer the interested reader to a recent study by Darrel Frost and his colleagues (see the "Bibliography"). The reader may choose to explore other perspectives on how species of frogs and toads should be grouped, some of which are presented by David Wake and his colleagues on AmphibiaWeb.

Distinguishing among species is essential for more reasons than just knowing about the biology of frogs. Most conservation initiatives are based on the conservation of imperiled species (e.g., U.S. Endangered Species Act), and, thus, knowing what groups of animals constitute individual species is vital for their conservation. However, one should remember that the real relationships among frog species and groups of species will not change substantially in our lifetime, and frogs likely do not care to which family any particular individual of *Homo sapiens* believes they belong.

Appendix A

Synonyms of Frog Scientific Names

The following names (column one) used in this book are traditional scientific names that have been proposed for change as a consequence of molecular studies that have resulted in re-evaluating the relationships among numerous amphibian genera and species. The proposed names, based on Frost (2010; see the "Bibliography"), are given in the second column.

Usage in book	Revision by Frost (2010)
Bufo alvarius	*Incilius alvarius*
Bufo americanus	*Anaxyrus americanus*
Bufo baxteri = hemiophrys	*Anaxyrus baxteri*
Bufo blombergi	*Rhaebo blombergi*
Bufo boreas	*Anaxyrus boreas*
Bufo coniferus	*Incilius coniferus*
Bufo fowleri	*Anaxyrus fowleri*
Bufo houstonensis	*Anaxyrus houstonensis*
Bufo latastii	*Pseudepidalea latastii*
Bufo marinus	*Rhinella marina*
Bufo microscaphus	*Anaxyrus microscaphus*
Bufo peltocephalus	*Peltophryne peltochephala*
Bufo periglenes	*Incilius periglenes*
Bufo punctatus	*Anaxyrus punctatus*
Bufo quercicus	*Anaxyrus quercicus*
Bufo speciosus	*Anaxyrus speciosus*
Bufo terrestris	*Anaxyrus terrestris*
Bufo viridis	*Pseudepidalea viridis*
Cyclorana novaehollandiae	*Litoria novaehollandiae*
Cyclorana platycephala	*Litoria platycephala*
Dendrobates pumilio	*Oophaga pumilio*
Rana capito	*Lithobates capito*
Rana capito sevosa	*Lithobates sevosus*
Rana catesbeiana	*Lithobates catesbeianus*
Rana chiricahuensis	*Lithobates chiricahuensis*
Rana clamitans	*Lithobates clamitans*
Rana grylio	*Lithobates grylio*

Rana heckscheri	*Lithobates heckscheri*
Rana palustris	*Lithobates palustris*
Rana pipiens	*Lithobates pipiens*
Rana ridibunda	*Pelophylax ridibundus*
Rana sphenocephala	*Lithobates sphenocephalus*
Rana sylvatica	*Lithobates sylvaticus*
Rana vitrea	*Amolops vitreus*

Appendix B

Resources, Organizations, and Societies for Frog and Toad Conservation

A diversity of regional, national, and international organizations and societies focus on frog conservation directly or contribute indirectly by publishing or supporting research that provides scientific data needed to make data-based decisions. Some organizations assist frog and toad conservation efforts by reporting information on conservation issues through electronic mailing lists, announcements, brochures, newsletters, and the popular press. The following list is not all inclusive but provides many of the groups that make significant contributions to frog conservation.

Reptile Zoos and Public Aquariums with Frogs and Toads in Exhibits

American Museum of Natural History, New York, NY
Atlanta Botanical Garden, Atlanta, GA
Bronx Zoo, Bronx, NY
Dallas Zoo, Dallas, TX
Detroit Zoo, Detroit, MI
Disney's Animal Kingdom, Orlando, FL
Edisto Island Serpentarium, Edisto Island, SC
Fort Worth Zoo, Fort Worth, TX
Gladys Porter Zoo, Brownsville, TX
Jacksonville Zoo, Jacksonville, FL
Houston Zoo, Houston, TX
Miami Metrozoo, Miami, FL
Munster Zoo, Münster, Germany
National Aquarium in Baltimore, Baltimore, MD
New England Aquarium, Boston, MA
Quito Zoo, Guayllabamba, Ecuador
Riverbanks Zoo and Garden, Columbia, SC
San Antonio Zoo, San Antonio, TX
San Diego Zoo, San Diego, CA
Shedd Aquarium, Chicago, IL
South Carolina Aquarium, Charleston, SC

St. Louis Zoo, St. Louis, MO
Steinhart Aquarium, San Francisco, CA
Tennessee Aquarium, Chattanooga, TN
Toronto Zoo, Toronto, Ontario, Canada
Zoo Atlanta, Atlanta, GA
Zoological Society of London (ZSL) London Zoo, Regents Park, London
Zoological Society of London (ZSL) Whipsnade Zoo, Whipsnade,
 Bedfordshire, England

Scientific and Conservation Societies That Identify Frogs and Toads as Part of Their Focus

American Society of Ichthyologists and Herpetologists (*Copeia*)
Australian Herpetological Society
Austrian Herpetological Society (*Herpetozoa*)
British Herpetological Society (*Herpetological Journal: Herpetological Bulletin*)
Brazilian Society of Herpetology (*South American Journal of Herpetology*)
Herpetological Association of Africa (*African Journal of Herpetology*)
The Herpetological Society of Japan
Herpetologists' League (*Herpetologica*)
International Society for the History and Bibliography of Herpetology (*Bibliotheca Herpetologica*)
New Zealand Herpetological Society
Partners in Amphibian and Reptile Conservation (PARC)
Society for the Study of Amphibians and Reptiles (*Journal of Herpetology*; *Herpetological Review*)
Society of European Herpetologists (*Amphibia-Reptilia*)

Magazines and Journals That Specialize in Frogs and Toads, Amphibians, or General Herpetology

Acta Zoologica
African Journal of Herpetology
Alytes
Amphibia-Reptilia
Bibliotheca Herpetologica
Bulletin of the American Museum of Natural History
Catalogue of American Amphibians and Reptiles
Copeia
Hamadryad
Herpetologica

Herpetological Conservation and Biology
The Herpetological Journal
Herpetological Review
Herpetozoa
Journal of Herpetology
Phyllomedusa
Reptile and Amphibian Magazine
Reptiles and Amphibians: Conservation and Natural History
Reptiles Magazine
South American Journal of Herpetology
Vivarium Magazine

Online Resources

Amphibian Ark, www.amphibianark.org

Amphibian Research and Monitoring Initiative (USGS), armi.usgs.gov

Amphibian Species of the World (American Museum of Natural History), research.amnh.org/vz/herpetology/amphibia

AmphibiaWeb (University of California), amphibiaweb.org

North American Amphibian Monitoring Program, www.pwrc.usgs.gov/naamp

Partners in Amphibian and Reptile Conservation, www.parcplace.org

Save the Frogs, www.savethefrogs.com

Bibliography

Alford, R. A., and S. J. Richards. 1999. Global amphibian declines: A problem in applied ecology. *Annual Review of Ecology and Systematics* 30:133–165.

Anderson, A. M., D. A. Hawkos, J. T. Anderson. 1999. Diet composition of three anurans from the playa wetlands of northwest Texas. *Copeia* 1999:515–520.

Bain, R. H., B. L. Stuart, and N. L. Orlov. 2006. Three new Indochinese species of cascade frogs (Amphibia: Ranidae) allied to *Rana archotaphus*. Copeia 2006: 43–59.

Bartlett, R. D., and P. P. Bartlett. 1998. *A Field Guide to Florida Reptiles and Amphibians*. Houston, TX: Gulf Publishing Company.

Beane, J. C., A. L. Braswell, J. C. Mitchell, W. M. Palmer, and J. R. Harrison III. 2010. *Amphibians and Reptiles of the Carolinas and Virginia*. 2nd ed. Chapel Hill: University of North Carolina Press.

Beltz, E. 2005. *Frogs: Inside Their Remarkable World*. Buffalo, NY: Firefly Books.

Blaustein, A. R., and B. Waldman. 1992. Kin recognition in anuran amphibians. *Animal Behavior* 44:207–221.

Brennan, T. C., and A. T. Holycross. 2006. *A Field Guide to the Amphibians and Reptiles in Arizona*. Phoenix: Arizona Game and Fish Department.

Bridges, A. S., and M. E. Dorcas. 2000. Temporal variation in anuran calling behavior: Implications for surveys and monitoring programs. *Copeia* 2000:587–592.

Caldwell, J. P. 1982. Disruptive selection: A tail color polymorphism in *Acris crepitans* in response to differential predation. *Canadian Journal of Zoology* 60:2812–2828.

Caldwell, J. P., and M. C. deAraújo. 1998. Cannibalistic interactions resulting from indiscriminate predatory behavior in tadpoles of poison frogs (Anura: Dendrobatidae). *Biotropica* 31:92–103.

Carroll, Robert. 2009. *The Rise of Amphibians: 365 Million Years of Evolution*. Baltimore: Johns Hopkins University Press.

Cogger, H. G. *Reptiles and Amphibians of Australia*. 6th ed. Sanibel, FL: Ralph Curtis Books.

Collins, J. T., and S. L. Collins. 1993. *Amphibians and Reptiles of Kansas*. 3rd ed. Lawrence: University of Kansas, Natural History Museum.

Conant, R., and J. T. Collins. 1991. *A Field Guide to Reptiles and Amphibians of Eastern and Central North America*. 3rd ed. Boston: Houghton Mifflin.

Crossland, M. R., and R. A. Alford. 1998. Evaluation of the toxicity of eggs, hatchlings and tadpoles of the introduced toad (*Bufo marinus*) (Anura: Bufonidae) to native Australian aquatic predators. *Austral Ecology* 23:129–137.

Crother, B. I., ed. 1999. *Caribbean Amphibians and Reptiles*. New York: Academic Press.

Da Silva, H. R., M. C. de Britto-Pereira, and V. Caramaschi. 1989. Frugivory and seed dispersal by *Hyla truncata*, a Neotropical tree frog. *Copeia* 1989:781–783.

Dayton, G. H., and L. A. Fitzgerald. 2001. Competition, predation, and the distributions of four desert anurans. *Oecologia* 129:430–435.

Degenhardt, W. G., C. W. Painter, and A. H. Price. 1996. *Amphibians and Reptiles of New Mexico*. Albuquerque: University of New Mexico Press.

Degraaf, R. M. 1983. *Amphibians and Reptiles of New England: Habitats and Natural History*. Amherst: University of Massachusetts Press.

Dodd, C. K., Jr. 2004. *The Amphibians of Great Smoky Mountains National Park*. Knoxville: University of Tennessee Press.

Dodd, C. K., Jr., ed. 2010. *Amphibian Ecology and Conservation: A Handbook of Techniques*. New York: Oxford University Press.

Dole, J. W. 1965. Summer movements of adult leopard frogs, *Rana pipiens schreber*, in Northern Michigan. *Ecology* 46:236–255.

Dorcas, M., and W. Gibbons. 2008. *Frogs and Toads of the Southeast*. Athens: University of Georgia Press.

Dorcas, M. E., S. J. Price, J. C Beane, and S. S. Cross. 2007. *The Frogs and Toads of North Carolina*. Raleigh: North Carolina Wildlife Resources Commission.

Drewry, G. E., and K. L. Jones. 1976. A new ovoviviparous frog, *Eleutherodactylus jasperi* (Amphibia, Anura, Leptodactylidae), from Puerto Rico. *Journal of Herpetology* 10:161–165.

Duellman, W. E., and L. Trueb. 1994. *Biology of Amphibians*. Baltimore: Johns Hopkins University Press.

Dundee, H. A., and D. A. Rossman. 1996. *Amphibians and Reptiles of Louisiana*. Baton Rouge: Louisiana State University Press.

Fahrig, L., J. H. Pedlar, S. E. Pope, P. D. Taylor, and J. E. Wegner. 1995. Effect of road traffic on amphibian density. *Biological Conservation* 73:177–182.

Federle, W., W. J. P. Barnes, W. Baumgartner, P. Drechler, and J. M. Smith. 2006. Wet but not slippery: Boundary friction in tree frog adhesive toe pads. *Journal of the Royal Society* 3:389–397.

Frost, D. R. 2009. Amphibian Species of the World: An Online Reference. Version 5.3, February 12, 2009. American Museum of Natural History, New York. http://research.amnh.org/herpetology/amphibia.

Gibbons, J. W., and R. D. Semlitsch. 1991. *Guide to the Reptiles and Amphibians of the Savannah River Site*. Athens: University of Georgia Press.

Gibbs, J. P., A. R. Breisch, P. K. Ducey, G. Johnson, J. Behler, and R. Bothner. 2007. *The Amphibians and Reptiles of New York State*. New York: Oxford University Press.

Gibson, R. C., and K. R. Buley. 2004. Maternal care and obligatory oophagy in *Leptodactylus fallax*: A new reproductive mode in frogs. *Copeia* 2004:128–135.

Glaw, R., and M. Vences. 1997. Anuran eye colouration: definitions, variation, taxonomic implications and possible functions. *Herpetologia1997 Bonnensis* 1997:125–138.

Gordon, M. S., and V. A. Tucker. 1965. Osmotic regulation in the tadpoles of the crab-eating frog (*Rana cancrivora*). *Journal of Experimental Biology* 42:437–445.

Greenberg, C. H., and G. W. Tanner. 2005. Spatial and temporal ecology of oak toads (*Bufo quercicus*) on a Florida landscape. *Herpetologica* 61:422–434.

Heyer, W. R., M. A. Donnelly, R. W. McDiarmid, L. C. Hayek, and M. S. Foster, eds. 1994. *Measuring and Monitoring Biological Diversity: Standard Methods for Amphibians*. Washington, DC: Smithsonian Institution Press.

Hoffman, E. A., and M. S. Blouin. 2000. A review of colour and pattern polymorphisms in anurans. *Biological Journal of the Linnean Society* 70:633–665.

Hutchison, V. H. 1998. The goliath frog (*Conraua goliath*): Physiological ecology of the largest anuran. International Symposium on Animal Adaptation (abstract), pp. 22–26.

Ishii, S., K. Kubokawa, M. Kikuchi, and H. Nishio. 1995. Orientation of the toad, *Bufo japonicus*, toward the breeding pond. *Zoological Science* 12:475–484.

Jensen, J. B., C. D. Camp, J. W. Gibbons, and M. J. Elliott, eds. 2008. *Amphibians and Reptiles of Georgia*. Athens: University of Georgia Press.

Kats, L. B., and R. G. VanDragt. 1986. Background color-matching in the spring peeper, *Hyla crucifer*. *Copeia* 1986:109–115.

Kerby, J. L., K. L. Richards-Hrdlicka, A. Storfer, and D. K. Skelly. 2010. An examination of amphibian sensitivity to environmental contaminants: Are amphibians poor canaries? *Ecology Letters* 13:60–67.

Kiesecker, J. M. 2002. Synergism between trematode infection and pesticide exposure: A link to amphibian limb deformities in nature? *Proceedings of the National Academy of Sciences of the United States of America* 99:9900–9904.

Kiesecker, J. M., and A. R. Blaustein. 1995. Synergism between UV-B radiation and a pathogen magnifies amphibian embryo mortality in nature. *Proceedings of the National Academy of Science* 92:11049–11052.

Lang, E., C. Gerhardt, and C. Davidson. 2009. *The Frogs and Toads of North America*. Boston: Houghton Mifflin Harcourt.

Lannoo, M., ed. 2005. *Amphibian Declines: The Conservation Status of United States Species*. Berkeley: University of California Press.

Lannoo, M. 2008. *Malformed Frogs: The Collapse of Aquatic Ecosystems*. Berkeley: University of California Press.

Lardner, B., and M. bin Lakim. 2002. Animal communication: Tree-hole frogs exploit resonance effects. *Nature* 420:475.

Lee, J. C. 2000. *A Field Guide to the Amphibians and Reptiles of the Maya World*. Ithaca, NY: Cornell University Press.

Lewis, E. R., P. M. Narins, K. A. Cortopassi, W. M. Yamada, E. H, Poinar, S. W. Moore, and X. Yu. 2001. Do male white-lipped frogs use seismic signals for intraspecific communication? *American Zoologist* 41:1185–1199.

Levine, R. P., J. A. Monroy, and E. L. Brainerd. 2004. Contribution of eye retraction to swallowing performance in the northern leopard frog, *Rana pipiens*. *Journal of Experimental Biology* 207:1361–1368.

Liner, E. 2005. *The Culinary Herpetologist*. Salt Lake City, UT: Bibliomania.

Loveridge, J. P., and P. C. Withers. 1981. Metabolism and water balance of active and cocooned African bullfrogs *Pyxicephalus adspersus*. *Physiological Zoology* 54:203–214.

Mammerson, G. A. 1999. *Amphibians and Reptiles of Colorado*. 2nd ed. Boulder: University of Colorado Press.

McCay, M. G. 2001. Aerodynamic stability and maneuverability of the gliding frog *Polypedates dennysi*. *Journal of Experimental Biology* 204:2817–2826.

McLeod, D. S. 2008. A new species of big-headed, fanged dicroglossine frog (genus *Limnonectes*) from Thailand. *Zootaxa* 1807:26–46.

Measey, G. J., and R. C. Tinsley. 1998. Feral *Xenopus laevis* in South Wales. *Herpetological Journal* 8:23–27.

Meshaka, W. E., Jr., B. P. Butterfield, and J. B. Hauge. 2005. *Exotic Amphibians and Reptiles of Florida*. Melbourne, FL: Krieger Publishing Company.

Minton, S. 2001. *Amphibians and Reptiles of Indiana*. Indianapolis: Indiana Academy of Science.

Mount, R. M. 1996. *The Reptiles and Amphibians of Alabama*. Tuscaloosa: University of Alabama Press.

Narins, P. M., G. Ehret, and J. Tautz. 1988. Accessory pathway for sound transfer in a Neotropical frog. *Proceedings of the National Academy of Science* 85:1508–1512.

Newman, R. A. 1988. Adaptive plasticity in development of *Scaphiopus couchii* tadpoles in desert ponds. *Evolution* 42:774–783.

Nussbaum, R. A., and E. D. Brodie. 1983. *Amphibians and Reptiles of the Pacific Northwest*. Moscow: University of Idaho Press.

Pittman, S. E., A. L. Jendrek, S. J. Price, and M. E. Dorcas. 2008. Habitat selection and site fidelity of Cope's gray treefrog (*Hyla chrysoscelis*) at the aquatic-terrestrial ecotone. *Journal of Herpetology* 42:378–385.

Rohr, J. R., and P. W. Crumrine. 2005. Effects of an herbicide and an insecticide on pond community structure and processes. *Ecological Applications* 15:1135–1147.

Ryan, M. J. 1985. *The Tungara Frog: A Study in Sexual Selection and Communication*. Chicago: University of Chicago Press.

Schiotz, A. 1999. *Treefrogs of Africa*. Frankfurt, Germany: Andreas S. Brahm.

Semlitsch, R. D., ed. 2003. *Amphibian Conservation*. Washington, DC: Smithsonian Institution Press.

Sessions, S. K. 2003. What is causing deformed amphibians? In Semlitsch.

Sheridan, Jennifer A., and Joanne F. Ocock. 2008. Parental care in *Chiromantis hansenae* (Anura: Rhacophoridae). *Copeia* 2008:733–736.

Simon, M. P. 1983. The ecology parental care in a terrestrial breeding frog from New Guinea. *Behavioral Ecology and Sociobiology* 14:61–67.

Slavens, F. L., and K. Slavens. 1999. *Reptiles and Amphibians in Captivity, Breeding Longevity and Inventory*. Seattle, WA: Slaveware.

Souder, W. 2000. *A Plague of Frogs: The Horrifying True Story*. New York: Hyperion Press.

Stebbins, R. C. 2003. *Western Reptiles and Amphibians*. 3rd ed. New York: Houghton-Mifflin Company.

Stebbins, R. C., and N. W. Cohen. 1995. *A Natural History of Amphibians*. Princeton, NJ: Princeton University Press.

Storey, K. B., and J. M. Storey. 1984. Biochemical adaptation for freezing tolerance in the wood frog *Rana sylvatica*. *Journal of Comparative Physiology B: Biochemical, Systemic, and Environmental Physiology* 155:29–36.

Tocque, K., and R. C. Tinsley. 1995. The population dynamics of a desert anuran, *Scaphiopus couchii*. *Australian Journal of Ecology* 20:376–384.

Toledo, L. F., and C. F. B. Haddad. 2009. Colors and some morphological traits as defensive mechanisms in anurans. *International Journal of Zoology* 2009:1–12.

Twain, M., and J. Paul. 1867. *The Celebrated Jumping Frog of Calaveras County, and Other Sketches*. New York: C. H. Webb.

Veeranagoudar, D. K., R. S. Radder, B. A. Shanbhag, and S. K. Saidapur. 2009. Jumping behavior of semiterrestrial tadpoles in *Indirana beddomii* (Günth): Relative importance of tail and body size. *Journal of Herpetology* 43:680–684.

Voyles, J., L. Berger, S. Young, R. Speare, R. Webb, J. Warner, D. Rudd, R. Campbell, and L. F. Skerratt. 2007. Electrolyte depletion and osmotic imbalance in amphibians with chytridiomycosis. *Diseases of Aquatic Organisms* 77:113–118.

Wake, M. H. 1993. Evolution of oviductal gestation in amphibians. *Journal of Experimental Zoology* 266:394–413.

Warkington, I. G., D. Bickford, N. S. Sodhi, and C. J. A. Bradshaw. 2009. Eating frogs to extinction. *Conservation Biology* 23:1056–1059.

Wells, K. D. 2007. *The Ecology and Behavior of Amphibians*. Chicago: University of Chicago Press.

Wickramasinghe, D. D., K. L. Oseen, and R. J. Wassersug. 2007. Ontogenetic changes in diet and intestinal morphology in semi-terrestrial tadpoles of *Nannophrys ceylonensis* (Dicroglossidae). *Copeia* 2007:1012–1018.

Wright, A. H., and A. A. Wright. 1995. *Handbook of Frogs and Toads of the United States and Canada*. Ithaca, NY: Comstock Publishing Company.

Zug, G. R., L. J. Vitt, and J. P. Caldwell. 2001. *Herpetology: An Introductory Biology of Amphibians and Reptiles*. 2nd ed. New York: Academic Press.

Index

Page references in italics refer to illustrations.

Index